The Cognitive Approach to Conscious Machines

The Cognitive Approach to Conscious Machines

Pentti O. Haikonen

ia

IMPRINT ACADEMIC

Published in the UK by Imprint Academic
PO Box 200, Exeter EX5 5YX, UK

Published in the USA by Imprint Academic
Philosophy Documentation Center
PO Box 7147, Charlottesville, VA 22906-7147, USA

ISBN 0 907845 42 8

A CIP catalogue record for this book is available from the
British Library and US Library of Congress

Contents

Preface

Ever since early childhood I have been strangely aware of the special puzzle of self and consciousness; one's existence as an individual with awareness of one's own existence here and now, separate from others and the environment. Why do I exist, what am I anyway, could I have been someone else? Later on instead of clarifying the mystery deepened and more questions arose. What is thinking? What happens in the brain when I have a thought or try to solve a problem? What is intelligence, could I become more intelligent? I guess that many people have asked and will be asking these same questions over and over again.

I got access to computers for the first time when I was studying at the Helsinki University of Technology in the late sixties. This opened up a new approach to the question of intelligence. A computer could be programmed to produce results that were achievable earlier only by human mental efforts. Inspired by this possibility I wrote a rather small program that was able to produce short stories, every time different ones. This program utilized simple grammar and a modest vocabulary. Due to the rather clever selection of the overall topic the main shortcoming of these stories was not always apparent — the totally missing plot. How do you program a computer to produce meaningful stories with plots? I soon realized that this was extremely difficult if not impossible with grammar and vocabulary only. Meanings of the words and some real world information would be needed; perhaps the computer would have to posses some kind of ability to understand.

The idea of machine understanding and eventually machine cognition and consciousness haunted me ever since. What is involved in thinking, understanding and consciousness? Could a thinking machine be built? Could I do it? What would it take? But alas, the time was not right, computers were not powerful enough to allow even modest experiments, integrated circuit technology was not advanced enough and most crucially, no employer was ready to finance this kind of research, this was clear without asking.

In the nineties things suddenly changed. I found myself in an avant-garde high tech company research center that wanted to push the limits of leading edge technologies and thus was willing and had the means to support research with applications far in the future. This was Nokia Research Center and I am indebted to the head of the research center, Dr. Juhani Kuusi and my immediate superior, Dr. Hannu Nieminen for the continuing possibility to work in this extremely interesting but also controversial field. During this work I was also able to write my doctoral thesis on the subject at the Helsinki University of Technology under the supervision of professor Raimo Sepponen. I defended this thesis successfully in 1999; my opponents were professor Kimmo Alho of the Helsinki University and professor Igor Aleksander of Imperial College, London. Thank you once more.

In this book I try to summarize the main concepts behind my research; the relevant background information provided by cognitive psychology and cognitive neuroscience, the philosophy and attempted explanations of consciousness, the outline of my design philosophy for cognitive machines and ultimately an explanation of consciousness within the framework of the proposed cognitive machine architecture.

Philosophers have the luxury of presenting their work as theoretical questions; the elegance of these questions is the hallmark of excellence. We engineers do not have this luxury. Our ideas that first appear as a design philosophy must face the acid test of practicality. My case is not a different one and thus my work includes also actual neuron group microchip development for the eventual implementation of cognitive machines along the lines that are outlined in this book. This work is being done in cooperation with Technical Research Centre of Finland, Centre for Microelectronics and the thanks go to Mr. Arto Rantala there.

I want to express my gratitude to professor Mikko Sams of Helsinki University of Technology, professor Kimmo Alho of University of Helsinki and my colleague Esa Erola for reading the manuscript and giving valuable advice.

Finally I also want to thank my wonderful wife Sinikka for support and patience and my son Pete, without whom my possibilities to observe evolving human cognition would have been very limited.

CHAPTER 1

INTRODUCTION

One of these days,
not very far away in the future
a machine may, after a thorough reflection,
reach the conclusion: "I think; therefore I am... immaterial".

Hardly ever has a technological pursuit had such philosophical importance as that of the quest for conscious machines. Our own consciousness and even our self-existence are still poorly understood. What should we think then about machine consciousness if we were ever able to create it?

Will machine consciousness be possible at all, will it be the greatest invention of mankind or will it only remain a science fiction dream? Is our consciousness unique, excluding by its special nature all reproduction by technical means known to man today? Or, could artificial consciousness nevertheless be possible? What would it take to design a conscious machine and how should we approach this task? These are the questions that are addressed in this book.

A prominent hallmark of consciousness is the awareness of one's thoughts. Thus a prerequisite for consciousness would seem to be the ability to think, to have mental content to be aware of. A system cannot be aware of its thoughts if it does not have any. Therefore in order to create machine consciousness we should first seek to create thinking machines.

Our thoughts appear to us as immaterial. This gives rise to the problem of mind–body interaction. Would a thinking machine perceive its thoughts as immaterial, too? Will we create a ghost in the machine if we succeed in the construction of thinking machinery?

Everybody knows how to think, but what is thinking actually? Is it the silent "inner speech" that keeps going on in our heads whenever we are awake? Is there a language of thought with its own grammatical rules? If so, then the brain would assemble thoughts according to the rules of that grammar. But then, what would be the difference between the brain and a computer that also operates by rules? Could we build a cognitive machine by reproducing the rules of the brain in a computer? Could we thus create a thinking robot with a mind of its own and would it also be conscious? The answer to these questions may not be the first thought that comes to mind.

What exactly are the rules of the brain? So far we have had only vague ideas that have not offered a straightforward basis for computational implementation. But, on the other hand, do we actually need to know the rules of the brain? After all what we want are the products of intelligence, the products of thinking. If the answers given by a computer and the brain to a question are the same then aren't the computer and the brain also equal? Thus we would only need to find out computational rules that produce the results of intelligence. This is the "same results" hypothesis that is behind much of the present-day discipline of Artificial Intelligence (AI).

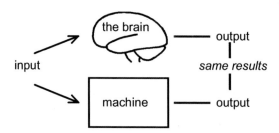

Fig. 1.1. A thinking machine via the "same results" principle

Computers are numerical machines. Sound and music are represented by strings of numbers describing the intensity of the sound at each moment. Likewise visual information is represented by the brightness values of picture elements. Consequently whatever is to be deduced from the auditory or visual information is achieved via numerical computation. Thus Artificial Intelligence calls for computational rules for each problem in the application. These rule sets, programmed strings of instructions are also called algorithms.

AI has imitated the products of intelligence by dedicated case-by-case algorithms. So far however, these algorithms have not constituted real intelligence outside their limited application, perhaps not even there.

The shortcomings of classical AI can also be attributed to the difficulty or implausibility of foreseeing and preprogramming each and every relevant rule of inference. Self-learning parallel networks, more or less modeled after biological neurons and synapses, have been proposed as a solution as with these no preprogrammed rules are needed. These artificial neural networks, also known as connectionist networks, adjust themselves to large sets of incoming data and produce output according to the implicit rules that they have acquired statistically. However, usually these rules cannot be easily recovered as formulas and the actual workings of a neural network may remain obscure to a human observer.

Traditionally neural computing has been applied to strictly defined problems such as pattern recognition, classification, categorization, function approximation, signal prediction, etc. More often than not neural computing is nowadays performed by a conventional computer so that the speed advantage of the inherent parallelism is lost and the actual process is reduced to a calculation style.

Artificial Intelligence and Neural Networks have produced some remarkable results, but these approaches do not excel in applications where true understanding is needed. Speech, text, story understanding, text summarization or image, scene, episode and movie understanding are especially difficult areas. The performance of programs in these fields has varied from barely acceptable to outright ridiculous. As for general intelligence and true creativity, there has been definitely none. At the end of the day no thinking machine has been built and the human brain still remains far superior to any computer.

So what is wrong here, why has the computational production of "same results" been so elusive?

Is the brain really a rule-based self-programmed computer? Does the brain execute programmed commands? Has Nature devised a program for everything we do, every motion, every emotion, every thought?

No, the brain is definitely not a computer. Thinking is not an execution of programmed strings of commands. The brain is not a numerical calculator either. We do not think by numbers. The brain does not represent its mental content by numbers. Therefore it should not be any wonder that apart some trivial cases "same results" cannot be easily achieved by computational algorithms.

The real sources of mind and intelligence are the non-numeric cognitive processes of the brain, and we should study and seek to reproduce these in order to achieve real machine cognition, comparable to that of humans. So, instead of the "same results" principle I am proposing here a "same processes" principle.

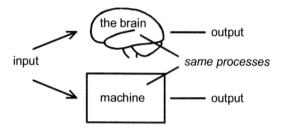

Fig.1.2. A thinking machine via the "same processes" principle

What is involved in the cognitive processing style of the brain? The human mind seems to operate with the flow of inner imagery, inner speech and sensations, with their meanings and significance. Cognition is also accompanied by feelings, emotions and consciousness. We can observe these, but with our natural means we cannot observe any possible material processes behind these. We are not able to introspect the actual neural firings inside the brain. Thus these processes are seemingly immaterial and therefore some people consider their implementation by material means impossible.

However, the seat of thinking and consciousness, the brain, is very much a material entity. Therefore it should be safe to maintain that a material thinking machine can be built because we already have one, the brain.

Thus we should first set out to study the processes of the brain. Neuroscience has found out that the brain uses stereotyped electrical signals for all processing and that the basic signal processing element is a biological cell, the neuron. Now we could start with the operation of the neuron, which is fairly easy to study, and proceed to investigate the combined operation of ever-larger neural assemblies until the whole brain is covered. This however, is a very tedious approach, it would be the same as to figure out how a computer works without any system-level knowledge, only by tracing the signal paths between the millions of transistors. This is not practical and therefore some information about the higher-level cognitive processes would be needed.

This kind of information is provided by cognitive psychology and cognitive neuroscience. The operation of our senses, eyes, ears, etc. is quite well known up to the initial processing in the brain. There are general theories for perception, attention, learning, memory, emotions, etc. The search for thinking machines is now transformed into the study of these cognitive processes and their possible implementation by electronic means.

In principle there are two different approaches available. We could try to create programs that reproduce cognitive processes directly at higher representational level or we could go down to signal level representations and create neural network systems on silicon. The first approach might seem to be more attractive as we could use existing computers for the job. However, there may be some fundamental problems here. We may ask if programmed cognitive processes really equal those of the brain: do we really get the real thing. A part of human cognition deals with representations, these we can easily implement in a computer. Human cognition involves also system reactions that are characterized by a specific feeling such as pain. These can also be represented in a computer but not truly realized because, as I see it, the computer does not have the respective system reactions, therefore it will not feel anything. And if we eventually could get it working it would still basically be a computer, subject to system errors and crashes.

I am proposing here the hard way, realization with novel artificial neurons and non-numeric signal-level representations that carry dedicated meanings. I will also propose a special cognitive architecture to reproduce the processes of perception, inner imagery, inner speech, pain, pleasure, emotions and the cognitive functions behind these. This machine would produce higher-level functions by the power of the elementary processing units, the artificial neurons, without algorithms or programs. However, there is a penalty; this machine cannot be implemented by existing digital microcircuits, therefore dedicated chips would have to be designed.

How about consciousness then? The proposed machine is a thinking machine that will be able to sustain the flow of inner imagery and inner speech, "mental content". The machine has also the faculties of perception and introspection that allow awareness of the environment and the machine's mental content. The real challenge of consciousness is its apparent immaterial nature. Does a "conscious" machine really perceive the flow of inner speech and imagery as immaterial? Does it perceive the possession of an immaterial self? Is it aware of its own existence? Answers will be proposed to these questions.

Part I of this book begins with a little bit of history and a review of computing principles. Arguments against the concept of thinking as the execution of strings of program commands are presented. Artificial neural networks have been proposed as a better way to realize thinking machines. Therefore the classical artificial neural network approach is also summarized. Again, arguments against neural network-based thinking machines are presented.

In part II I discuss the fundamentals of cognition and consciousness, the stuff that should be implemented in any real cognitive machine worth its while.

In part III I present my design philosophy and model for machine cognition and consciousness and discuss the general nature of consciousness in the light of this model.

PART I

THINKING AND COMPUTATION

CHAPTER 2

GOOD OLD-FASHIONED ARTIFICIAL INTELLIGENCE

Thinking Machines, 40s Style

Imagine a large hall full of tall metal cabinets, with panels of thousands of electron tubes visible. This is a machine that you have never seen before, a behemoth of high technology. Put some data in and in no time at all this machine will perform calculations so complex that you could not complete with pen and paper for months, calculations that you might not even master at all. What is this machine doing, is complex mathematics not thinking? You are impressed by the sheer size of the machine, you don't understand what is going on. You see the romantic glow of electron tubes in darkness, you hear the machine humming, you could so easily believe it's alive. This is also a machine that you will never see again; this was the Zeppelin of information technology, this was the computer of the forties.

It is no wonder that awe of the new technology led prominent scientists ask the questions: Does a computer think? What is thinking actually? A computer operates by following logical steps written in a computer language. If we accept that thinking is the silent speech in our heads and this speech is composed according to linguistic rules and logic then obviously thinking, the language of thought, is not different from computer languages. The point is, if results that are equivalent to those achieved by thinking can only be achieved through computations, that is, by program execution, there is no other explanation. Therefore thinking is computation. So, in principle any logically consistent thought could be produced by a computer program. Logically inconsistent thoughts or emotions — good riddance, who would really need those? Therefore, running a program is equivalent to thinking even if in a strange language. The philosophy of Artificial Intelligence (AI) was born.

In 1958 Nobel laureate and AI pioneer Herbert Simon said: "...there are now in the world machines that think that learn and create.... in visible future the range of problems they can handle will be coextensive with the range to which the human mind has been applied"[31]. Yes, the computer is here and it is the thinking machine, even the worst skeptic will realize this as soon as the machines get bigger and faster and we have better programs. This was more or less the euphoric feeling in those days and is not totally unknown even today.

Symbols and Computers

Diamonds are symbols for eternal love and the signs of the Zodiac symbolize the supposed fates of people. Road signs specify speed limits and forbidden directions. Flowers to a lady, candle light, they all symbolize something. A symbol has a meaning and significance. Art, advertising, interpersonal communication, in fact most of our everyday life is based on the use of symbols.

Computers operate by processing symbols, too, but make no mistake here, the ones above are not the symbols that computers use. Instead in strict computing sense symbols are only marks, usually strings of "ones" and "zeros" that can be recognized by their appearance. There are no meanings attached and no processing is based on meanings. Computers utilize sets of rules that specify how these symbols are to be manipulated, what symbol is to follow the given symbols in each instance. Basically a computer accepts strings of symbols as the input and delivers other strings of symbols as the output. It remains for the human operator to attach meanings to these symbol strings.

However, even though the computer itself does not operate with the meanings of the symbols, meanings may yet be attached by peripheral machinery instead of humans, there is no fundamental objection to that. For instance a computer may be used to guide the motion of a robot. In that case the meaning of the output symbols would be the motion to be executed and the meaning of the input symbols would be the desired action and the measurements of the present position of the robot.

So, *symbolic computing is rule-based manipulation of symbols that do not have any attached meanings or significance.* A set of output symbols is computed by rules from a given set of input symbols.

Computing by Look-up Tables

The simplest computing device might be a black box with inputs and outputs with rules inside that determine the proper output for each input. These internal rules could be implemented by combinatorial logic, like "inputs A AND B give output C". Or, instead of any logic circuitry, there might be a table where the proper output for each possible input would be listed, a look-up table. From outside the operation of the black box would be the same in each case, we would have a stimulus–response system. This is also a practical device. Today these, even very large ones, can be realized either as combinatorial logic circuitry or look-up tables consisting of read-only memories.

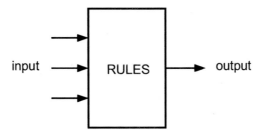

Fig. 2.1. The black box computing device

What could be computed with a system that is this simple? There is a surprisingly large wealth of tasks that can be performed. We can do transforms between representations like from Fahrenheit to Centigrade, from millimeters to inches, etc. We can do pattern recognition and classification by making given output signals represent distinct sets of input signals.

Could we compute everything with a large enough look-up table? Would it be possible, at least in principle, that for a given application we could compute everything in advance and create a large look-up table, regardless of the complexity of required computations?

Suppose now that the input to the black box would consist of eight signals, each having a possible value of one or zero. Then there would be $2^8 = 256$ different input combinations to be listed in the look-up table, not too difficult. But, in practice we might need perhaps 32, 64 or even thousands of input signals. For instance, 2^{32} would already give 4 294 967 300 combinations, that is about 4.3 Gigabits in computer terms. We can see that very soon the size of the look-up table would be much too large to be actually implemented as a read-only memory on a microchip. How do we compress the table?

Instead of using a list of ready-computed outputs we can use combinatorial logic circuitry that computes the proper output every time when inputs are presented. This circuitry would implement the rule that specifies the relation between the inputs and outputs. If this circuitry could be implemented with less silicon than the respective read-only memory look-up table then we would have achieved the desired compression. Usually this is the case. Very large tables can be reduced into computations with simple rules. The actual compression depends on the complicity of the input–output relation. If each input–output pair required a rule of its own then no compression would be possible and the read-only memory look-up table would be the most compact realization. In this case the output would seem to be in random relation to the input as no universal rule could be devised. Random data cannot be compressed, a fact that is known to those of us with experience in audio and video compression.

Nevertheless, for theoretical purposes we can consider the black box model as a look-up table as the internal realization of the box has only practical consequences.

Next we would like to ask, what are the limits of this approach? Could we implement every possible computation as a look-up memory operation, could we squeeze a PC into a very, very large look-up table?

The basic "stimulus–response" style look-up table approach has a fundamental deficiency that sets limits to the executable computations. The output is a function of the instantaneous inputs only, it does not depend on the history, the past inputs or outputs. Therefore the look-up table or the equivalent combinatorial logic is not able, for example, to classify and recognize input sequences or to produce output sequences if the inputs are removed. Much of computing is related to sequence processing, therefore we cannot replace a PC with a look-up table of any size.

Finite State Machines

Suppose that you are in a foreign city and want to find your way to the railway station. A friendly police officer gives you the advice: "At the next corner turn left and you are there". Now this is your look-up table that contains only one rule to be memorized: "Input; I am at the corner — output; must turn left". This is simple enough, but suppose now that you are further away from the railway station. The advice given by the police officer could now be: "At the next corner turn left, then at the next corner right, then right again and then left". Obviously there are now four rules to be memorized, each of the same form:

"Input; corner–output; turn direction". In addition to the actual rules you must also memorize their order: "Corner one; left, corner two; right, corner three; right, corner four; left". When you proceed towards the railway station you must keep count of the corners, having used the rule number one you must set the corner number to "two" and so on. Now in fact the rules have the form: "Inputs; corner, corner number — outputs; turn direction, next corner number".

What we have here is a look-up table with multiple sets of rules, the valid set being determined by an additional input value, the state of the look-up table system. The rules determine the actual output and also the next state for the system.

We can now visualize the system as a system with a finite number of rule sets and a state value determining the valid set. This system is known as the finite state machine, as the number of the possible rule sets and states is finite.

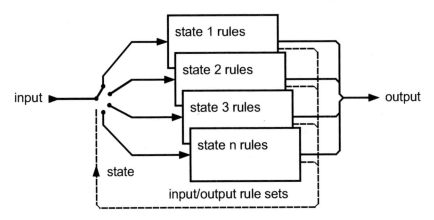

Fig. 2.2. The finite state machine as a system of rule sets

The finite state machine is usually drawn in the following way:

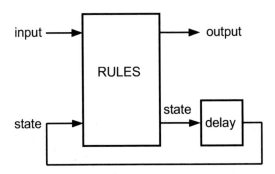

Fig. 2.3. The finite state machine

Thus a finite state machine has finite sets of input and output symbols, a finite set of internal states and internal rules that specify the next output and next state for the present input symbol and present state. The delay unit or other storage element in the feedback loop is needed to sustain the present state long enough so that the next state will only have effect at the next input.

The finite state machine can be used to (1) generate sequences, (2) classify or recognize input sequences and (3) to map input sequences into output sequences.

A simple example of sequence generation by a finite state machine is the digital clock. A clock that shows minutes and hours must have 60 x 24 = 1440 states. The input is the clock pulse occurring once every minute. At each clock pulse the system changes its state according to the state table below. A unique display value, the actual output, is attached to each state to indicate the instantaneous time.

Table 2.1. State table for a finite state machine clock

input	state	next state	output
0	0	0	00:00
1	0	1	00:01
0	1	1	00:01
1	1	2	00:02
⋮	⋮	⋮	⋮
0	1439	1439	23:59
1	1439	0	00:00

We can also draw a state diagram where each state is represented by a circle and the transition from one state to another is represented by an arrow. A symbol attached to each arrow represents the respective input.

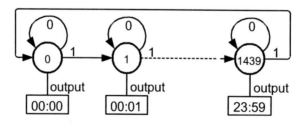

Fig. 2.4. The state diagram for a finite state machine clock

The finite state machine can be used to recognize input sequences. In this application the finite state machine is sometimes called the acceptor, or the Moore machine[60]. In the following example the state diagram is given for a finite state machine that recognizes the words "boot", "book" and "board".

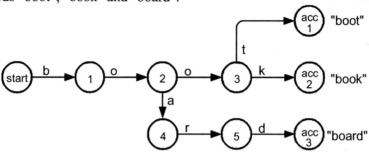

Fig. 2.5. The state diagram for a finite state machine word recognizer

In this example words are entered letter by letter beginning from the start-state. The machine goes through a sequence of states determined by the letters of the word. The word is recognized when the unique accept-state for that word is reached. Obviously this process will not tolerate any errors, which is good if secret passwords are to be detected but bad if speech is to be recognized.

Yet another frequent application for the finite state machine is the mapping of sequences into others. In this application the finite state machine is sometimes called the transducer or the Mealy machine[58]. Typical applications are for example bar code to binary translation, digital filters, edge detectors etc.

The history of the finite state machine affects its present state. However, the finite state machine is not able to store and retrieve intermediate results, therefore for instance calculations that involve multiple parentheses are not possible. We could make the machine more powerful if we only could make it store and recall intermediate results and other data as needed. Therefore the finite state machine would need a… need a what? If you had answered this difficult question back in the thirties, you would be famous now.

The Turing Machine

Alan Turing answered this question in 1936 by adding read-write memory to the finite state machine. And it was not a small amount of memory, it was infinite. This memory was realized by a moving infinite length tape and a head that could read, write and erase marks on that tape. The rule set was augmented by rules that controlled the movement of the tape and the read/write/erase operation. The tape itself contained the input data to the machine, intermediate results and eventually, if everything went right, the final answer. All this data was in the form of ones and zeroes with possible blanks in between. It took an operator or another machine to translate these into real numbers or alphabets. This hypothetical machine was to be known as the Turing Machine[105].

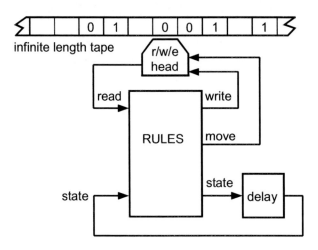

Fig. 2.6. The Turing machine

In fig. 2.6. the read/write/erase head reads a symbol from the tape. This symbol and the present state are the inputs to the machine. There are three outputs, the symbol to be written, tape move command and the next state symbol. A state table like the one below defines the operation of the machine.

Table 2.1. An example of a set of rules for a Turing machine

INPUTS		OUTPUTS		
state	read	write	move	next state
S1	blank	0	left	S2
	0	blank	right	S2
	1	1	left	S1
S2	blank	1	right	S2
	0	0	right	S2
	1	blank	left	S1

Two steps are needed to make the Turing Machine to solve a given problem. First a suitable state table (the rules) must be designed and then the initial values for the calculation must be given as symbols on the tape. Thereafter the machine will read the initial values and continue to change states and read/print symbols according to the state table and stop when the final result has been printed. A different state table is usually needed for each different task. The hypothetical Universal Turing Machine is however able to adjust its state table so that it can compute everything that any other given Turing machine can.

It should be obvious that a real Turing Machine cannot be built, as there is no way to realize the infinite length tape. In practice infinite memory is not needed and finite length tape will do fine. Thus limited length tape versions can be built and even more easily simulated by a computer program. However, the design of the state table for more complicated tasks may not exactly be an easy or especially pleasant exercise. The Turing Machine was not intended to be a practical computing device, instead it was a hypothetical model for computation theory.

Von Neumann Architecture

The Turing Machine architecture has nothing to do with present day computers. The Oscar here goes to Presper Eckert and John Mauchly who in 1945 outlined a stored program machine that later was to be known perhaps wrongly as the von Neumann computer architecture[73].

A computer with von Neumann architecture consists of three main units, namely a control unit, an arithmetic/logic unit and random access memory.

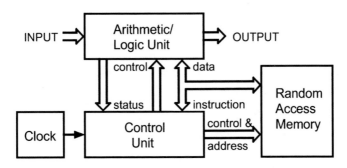

Fig. 2.7. The von Neumann computer architecture

The program, a string of instructions, is stored in the random access memory. The control unit reads one instruction at a time and commands the arithmetic/logic unit to perform the commanded action on the data. This action may be an arithmetic operation or data transfer between registers and memory locations, etc. A clock is used to time the operation so that one or a few instructions will be executed at each clock period. Data and instructions are represented by binary words, that is, arrays of binary ones and zeroes. The length of the binary word is usually eight, sixteen, thirty-two or sixty-four bits or even more.

The von Neumann computer has the essential elements of the Turing Machine; the random access memory and the equivalent of state machines embedded in the control unit and the arithmetic/logic unit. However, the architecture is far more practical and the programming easier as the basic instruction set can contain a large number of practical commands instead of the Turing Machine's awkward read, write and move commands.

The Turing and von Neumann machines are serial computers. They execute one instruction or command at a time no matter how much data might be available. This limitation is sometimes called the von Neumann bottleneck and is the reason for the need of ever-increasing clock frequencies to enhance the computing speed.

The von Neumann architecture is not the only practical computing architecture. Computing speed can also be increased by executing instructions in parallel. Also special digital signal processing architectures that are optimized for the calculation of given groups of algorithms exist. These are faster than comparable von Neumann machines, but even they are not able to compute anything that a Universal Turing Machine could not compute.

The Church-Turing Thesis

An algorithm is a string of commands by which a mathematical function is computed. Turing stated that any computable algorithm could be computed by a specific Turing Machine with a suitable set of rules. This is not a provable fact but rather a definition of computability. Then Turing went on to show that there exists a Universal Turing Machine that can compute everything that any specific Turing Machine can compute. Therefore, if there were a specific Turing Machine for every computable algorithm and these machines could be replaced by the Universal Turing Machine then the Universal Turing Machine could compute everything that was computable. This argument is also known as the Church-Turing thesis[18]. This thesis cannot be proven but could be disproved by finding an algorithm that cannot be computed by any Turing Machine. So far that kind of algorithm has not been found.

Digital computers compute by algorithms. According to the Church-Turing thesis the Universal Turing Machine can compute every computable algorithm. Therefore it can also compute every algorithm that any digital computer can. Thus there cannot exist another algorithmic computing machine that could compute something that the Universal Turing Machine could not compute. Thus the Church-Turing thesis leads to the claim that no computer can compute more than the Universal Turing Machine can. However, the thesis does not lead to the claim that no computer could be faster or be more practical than the Universal Turing Machine. In fact usually this is the case. It can be easily seen that it would be most awkward to use a Turing machine for any modern application like word processing, multimedia, computer games, or whatever.

The final lesson here is that machinery that is supposed to be able to compute every computable algorithm must have at least the equivalent power of a state machine combined with read/write memory. Machine architectures that amount to no more than look-up tables or state machines do not posses full computational power. This is a useful fact to remember whenever some "novel and revolutionary" computational schemes are to be evaluated.

The Great Imitation Game

The Church-Turing thesis leads to the claim that the Universal Turing Machine can compute everything that any other computer can do. If thinking were algorithmic computing then the brain would be a computer and then according to the Church-Turing thesis the Universal Turing Machine would also be able to compute whatever algorithms the brain would compute. Therefore the Universal Turing Machine would be able to think. As Universal Turing Machines are not readily available we must use the computers that we have. Thus the question is, can we program an existing computer so that the execution of the program would amount to thinking? And, if we had such a program, how could we determine whether the computer was actually thinking?

Do you think? I have no way to tell what happens inside your brain, but I can test you by asking questions. All humans are biologically more or less similar so if your answers were similar to those that I could give, then I would have a reason to believe that the mental processes inside your brain were also similar to mine and if I think then you think too.

Alan Turing thought that somewhat similar reasoning could be used to test if a computer can be said to think[106]. Turing's reasoning went like this: Suppose that there are three players, a man, a woman and an interrogator of either sex. The interrogator stays in a separate room and is not able to see the other two players. The only communication device between the interrogator and the two others is a computer screen and keyboard. The object of the game for the interrogator is to determine which one of the two is the man and which one is the woman. The interrogator is supposed to put questions to the man and woman and to determine by the answers which one is which. Now the man's object is to cause the interrogator to make the wrong identification by pretending to be a woman. Turing continues: "What will happen when a machine takes the part of the man in this game? Will the interrogator decide wrongly as often when the game is played like this as he does when the game is played between a man and a woman? These questions replace our original, can machines think."

It should be eminently clear that Turing's imitation game test is badly flawed. If the game is played with a real man and a woman and the man succeeds in convincing the interrogator that he is the woman then does it follow that the man is indeed a woman? Of course not. So what is the point here? The man-woman part of this test is totally irrelevant. Later followers of Turing have streamlined the test so that only one test subject exists, either a machine or a human. In this case it is the object of the machine to make the interrogator believe that he is dealing with a real thinking human.

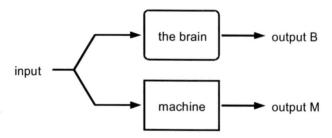

Fig. 2.8. The revised Turing test. If the brain and the machine give identical outputs for the same input, does it follow that the machine thinks?

At first glance this version of the test may look more rational, but it is still flawed. If you believe that you are dealing with the real thing does it necessarily follow that the thing is indeed real? Of course not. The computer and the human brain are not structurally similar so even if the computer succeeds, there is no reason to conclude that the result would be due to similar mental processes; thinking. The Turing test can be great fun, though.

Artificial Minds via Simulation?

Nowadays it should no longer be fashionable to equate the "language of thought" to computer languages. It should be clear to everybody that the running of a computer program does not constitute thinking, not even in a strange language. Thinking is definitely not an execution of the list of commands of a computer program. Instead, now AI people like to say that computers do the same things as the brain and programs that simulate human cognition can eventually surpass general human intelligence. It is true that AI programs in the fifties, sixties, seventies and even eighties did not quite deliver what was promised but then the computers were not very fast and their memory capacity was very limited.

Present day computers are very fast and their memory capacity is practically unlimited. Therefore computers can be used to model and simulate various complicated processes. Electronic circuit simulation is one common application. The circuit diagram of the device is simply drawn on the computer screen. Thereafter the simulation program computes the voltages and currents for each node in the circuit for any given inputs. Now the virtual circuit behaves as if it were assembled from real components and produces the same temporally varying values for voltages and currents as the real circuit would do. It is easier and faster to create virtual circuits instead of real ones and modify them if they do not work. Therefore circuit simulation is nowadays an indispensable tool for all electronic equipment design and especially for microcircuit development. Complicated circuits containing millions of transistors can be reliably simulated in this way. This kind of simulation does not have to take place in real time, the execution of the simulated function of a circuit may take much longer than with the actual hardware circuit. Sometimes however, the execution speed may be the same. In that case the actual hardware circuit can be replaced by the computer simulation, for instance a microcircuit that is being developed may be replaced in a circuit board by wired-in simulating computer. In this case the wired-in computer is called an emulator. The success of electronic circuit simulation is due the fact that the physical laws of electronics are well-known, therefore the behavior of electronic components can be accurately modeled, their input–output relationships, even highly non-linear ones can be accurately computed.

True, the computer is not the brain, but perhaps we could simulate thinking processes with it. In principle two approaches are possible. We can have the top-down approach where we model and define higher level cognitive functions and devise a program that realizes these. These higher level functions could be like the perception of environment, response generation by reactive processes, deliberative processes for planning and scheduling and overall controlling meta-management processes. The other way of simulation would be the bottom-up approach. This method would involve the modeling of actual neurons and their connections and building higher level functions from these.

It has been argued that bottom-up simulation of cognition is impossible because the gap between the basic operation of neurons and real cognitive functions cannot be bridged. This gap is similar to the difference between the transistors of a computer and a computer program. There is nothing in the transistors of a computer that determine what kinds of programs we run, whether we do serious computing or play computer games. There is also no way that we could deduce how a computer game works from the circuit diagram of the computer. This is because there is practically an infinite number of states and state sequences that the computer can generate and only a certain set of these sequences correspond to the running of a given program. Therefore, even if we were able to devise a "circuit diagram" for the brain, there would be nothing in the circuit to suggest any higher cognitive processes, it would not work. No combination of neurons and their firings would explain higher cognition, which could only be described at the system level, that is, by the methods of cognitive psychology and without any reference to the workings of individual neurons or neuron groups. Therefore cognitive processes could not be simulated by simulating the workings of individual neurons and neuron groups because the gap between the operation of neurons and the higher level cognitive processes could not be bridged. And as the gap cannot be bridged, neither can there be a way to design a cognitive system starting from neurons. This chain of arguments is false.

It is true that the circuit diagram of a computer does not explain the workings of a computer game. But, as soon as the game program has been loaded into the computer's memory, things change. Now it will be possible to trace out what will happen. We will find out that this is a game and not for instance a word processor. We will also understand generally what the game will do. In practice this may be quite tedious though, as it may amount to running the embedded program with pen and paper, inspecting the action of each transistor and gate, but that is not the point. There will be no missing bridge between the workings of the individual transistors and the actual high level program. Yes, we know that computers and computer games can be designed and their workings can be simulated by another computer.

Likewise cognitive systems can be designed by starting from individual neurons and their operation can be simulated — in principle. However, considering the immense number of neurons needed, this would seem to be a gargantuan task beyond the grasp of the human mind. How could we design systems that are this complicated?

Engineering has solved this problem already long ago. Complicated systems can be handled and designed by starting with the basic processing units, transistors, neurons, what ever, and assembling larger units whose operation can be explained by the operation of their individual components. Then these in turn can be combined into even larger units. The operation of these larger units can now be explained by the operation of their component units. Here only the input–output relationships of the component units need to be considered, reference to individual transistors or neurons is no longer necessary. This process can be continued by assembling ever-larger units in the similar way. At each level the input–output relationships of the previous level units only need to be considered. In this way systems of arbitrary complexity can be mastered.

The logic circuits of a computer consist of individual transistors. To understand the operation of a transistor you would need to know and understand solid state physics, electrons, energy levels of individual atoms, etc. However, a computer designer does not have to consider any of this. The understanding of the input–output relationships of the logic circuits is all that is needed for the design of a system, a working computer. The block diagram for a computer can be the same whether the logic circuits were assembled from relays, electron tubes, transistors or the like. True, some implementations are more practical than others, but that is not the point.

The argument that cognition can only be explained at system level, without any reference to the workings of neurons, leads now to another conclusion, perhaps to one that is rather unexpected or even contradicts the expectations of the perpetrators of this argument. If cognition can be explained without reference to the actual workings of individual neurons then the actual realization of the neuron is of no consequence — biological neurons can be replaced by artificial neurons with the same input–output characteristics!

However, there is a completely different reason why the creation of a thinking machine by using a computer as a bottom-up brain simulator is a dead-end street. The brain has a very large number of neurons and even larger number of synapses. The workings of every synapse would have to be computed in the simulation. Now if we assume the number of synapses to be 10^{14} and the simulation of one synapse to take ten clock cycles then with a 1 GHz processor one simulation cycle would take 10^6 seconds or 277 hours! The human brain does the same in a fraction of a second due to its parallel way of operation. It can be seen that no present or near future serially operating computer can do this. However, this fact does not exclude small-scale non-real-time simulation experiments with a reasonable number of neurons.

The top-down approach overcomes the speed problem by defining higher level cognitive functions directly, without any underlying elementary processes. Therefore the computational burden can be relieved and the execution can be faster.

A representative example of contemporary approaches to AI is professor Aaron Sloman's Cogaff program architecture standard model at Birmingham University[93]. According to this model the system consists of groups of programs responsible of the perception of environment, response generation by reactive processes, planning and scheduling via deliberative processes and overall control by meta-management processes. The system utilizes also variable threshold attentional filtering, long-term memory, motivational subprograms and alarms.

Arguments against AI

Many arguments have been put forward to show that a programmed computer does not and will not ever have the cognitive powers equivalent to human thinking. Some of these objections are summarized here.

The Gödel objection: In 1931 mathematician Kurt Gödel proved and published his famous theorem about arithmetic: Some arithmetic truths are not provable within the arithmetic system[38]. This theorem also means that there are true arithmetic statements that cannot be derived from arithmetic axioms by arithmetic rules. Therefore no Turing Machine can print all true statements of arithmetic. However, mathematicians are able to somehow see the obvious truth of these kinds of arithmetic statements.

The Gödel theorem has been taken to prove that computers can never equal human brains and therefore no artificial cognition is possible. Is this really so? How do the mathematicians themselves find these kinds of sentences? The Gödel theorem states that these sentences cannot be derived from the arithmetic axioms by the rules, so obviously the mathematicians are not following the rules of the book either, they are cheating. How are mathematicians able to "see" that a sentence would be true even if it could not be proven to be true or false? Would it be that the sentence could be "interpreted" by attaching external physical meanings to it and seeing that within that framework the sentence could be true? If this were the case then the Gödel theorem would only show that syntax, the rules only, are not always sufficient, some meanings must also be considered.

The argument against thinking machines by determinism: (1) Thinking is not deterministic, we have freedom of thought, we can think any thought as we please. (2) Machines and programs are deterministic, identical inputs and initial states will produce identical outputs every time. Machines do not have freedom of thought, therefore machines cannot think.

The "proof" for the first part of the argument, the non-deterministic nature of thought, is based on everyday experience. Firstly, we seem to be able to think about anything as we please. Secondly, if the same input or stimulus is given to us repeatedly, our brains will produce a different and to some extent unpredictable outcome each time. Is this real proof of the non-deterministic nature of the thought? Perhaps not. The freedom of thought may be illusory as the complex but deterministic causes of each thought may be beyond the reach of our consciousness.

Secondly, it is not really possible to give repeatedly exactly same stimuli to our brains to test determinism. Even if we were able to control completely the environment, and to give only the desired stimuli, the situation would not be the same. We cannot keep the temperature and chemical state of the brain constant. These affect the neural activity and will cause a different response each time. Moreover, the previous test run would also affect the brain, therefore we could not secure the same initial brain state for each of the test runs. So, it is not possible to test whether the brain would actually produce the same response for the same input and same initial brain state. Therefore we cannot experimentally prove that the brain is not deterministic.

A "thinking machine" of the kind that is considered here is coupled to the environment via sensors just like the human brain and therefore we must consider the environment and the machine as a combined system. In this kind of a system the inherent determinism of the machine is subdued even if one random variable is introduced. The environment, its objects, happenings, physical properties like temperature, humidity, etc. introduce enough randomness to suppress determinism in the machine–environment system. Obviously, this would apply to brain–environment systems as well, so the question about the inherent determinism of the machine or the brain is not really a relevant issue here. Therefore, the argument against thinking machines by determinism is not valid.

Computers do not understand. In the eighties philosopher John Searle devised a thought experiment that was intended to show that computers or computer programs are not capable of understanding[89,90,91]. This thought experiment went like this. Suppose that we have a closed room. Inside this room we have a person that does not understand Chinese. In the room there is also a large rulebook. Questions that are written in Chinese are forwarded to the person inside through a small hole in the wall and through this hole this person is supposed to deliver a correct answer, also in Chinese. Remember, this person does not understand Chinese, but he can compile the answer mechanically with the aid of the rulebook. Chinese questions are forwarded into the room and after a while perfect Chinese answers are coming back. The person inside does not understand Chinese, but does the room as a system understand?

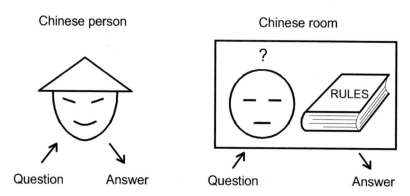

Fig.2.9. Searle's Chinese room. A Chinese person understands Chinese, but if the Chinese room gives the same answers to the same questions, does it follow that the room understands Chinese?

Searle compares the Chinese room to a computer. In the Chinese room the answers are compiled by following the rules in the book: "Chinese characters like this and in this order must be followed by these characters in this order." Searle reckons that a computer executes its program in the very same way, by syntax and with no reference to external meanings. By common sense the Chinese room does not understand Chinese, if we accept that then we must also accept that the computer does not understand anything. Searle says: "syntax is not sufficient for semantics" that is, rules do not convey meaning. This happens to be true by definition, no thought experiments would be needed for this.

But alas, the Chinese room thought experiment is very badly flawed. It should be immediately obvious that *the Chinese room is not a computer at all*, it is not even a state machine, instead it is merely a low-down look-up table falling short of any Turing power. Therefore this is not the analogy that Searle obviously intended, this does not constitute an argument against computers even if it were otherwise logically sound. At most the Chinese room metaphor could be used as an argument against understanding by look-up tables, but this is already trivial.

The shortcomings of the Chinese room are easy to see. Can we compile a correct answer to every question by rules only? Suppose we asked, "what is the temperature inside the room?" and only the true answer would be accepted. In this case the rulebook would be worthless unless it said in English (if the person inside was English speaking) that at this point the thermometer would have to be read and the reading would have to be translated in Chinese according to the given list. If we allow this then the room would no longer be a mere look-up table and as a system it would be definitely understanding something because the meaning of the given Chinese characters were being grounded to external world and action.

Now you may say: Yes, yes, but this is not the same kind of understanding that we humans have, real understanding is broader than this. This is true, human understanding involves complicated associative connections, imagery, values, etc. but this is already beside the point. Even present day computers do not operate without any reference to outside world, be it even their hardware, meanings do affect their operation. But, on the other hand it is quite safe to say that computers do not presently have inner imagery and the other mental stuff that we humans have.

Computers lack cognition. Computers are not creative. They only execute commands, they do not do anything of their own, they do not create anything new. All computer creations are products of the programmer. Computers do not have sudden insights. Computers do not have free will. Computers do not feel or have emotions and have no needs or drives. Computers do not have mental life, they do not operate with inner speech or inner imagery. This argument is definitely true for present day computers.

The non-feasibility of programming. Who can envision and program the computer for each possible situation? On the fourth of June 1996 the maiden flight of Ariane 5 rocket, the hope and pride of the European Space Agency, ended in failure and shame. Only 37 seconds after lift-off the rocket veered off, disintegrated and exploded. The shame: Guidance computers crashed because of a trivial software error that allowed the overflow of a register. It is easy to protect a program against this type of register overflow, yet here it was not done because the programmers believed that no overflow could possibly take place.

Who of us has not experienced the situation where a computer program, perhaps Microsoft Word, crashes, leaving us with a totally non-functional machine only to be resurrected by booting? Files that we have been working on for days are gone for good. A slight bug, a programming error and everything is lost. "These are not bugs, these are features" — what a ridiculous excuse. One can really wonder if intelligence can be programmed at all as long as the programmers themselves seem use it far too sparingly.

If programmed intelligence is not really feasible, then wouldn't it be much better to create machines without programs, machines that learn by themselves like children do? Artificial neural networks would seem to offer that possibility.

CHAPTER 3

ARTIFICIAL NEURAL NETWORKS TO THE RESCUE

The Biological Neuron

The human brain is a huge network of biological signal processing units, neurons. If we could model these neurons and the way in which they are cross-connected then we might be able to create similarly functioning networks with artificial, perhaps silicon-based, neurons. The human brain operates without programming, therefore in order to get rid of programs we should seek to copy the operational principles of biological neurons and the human brain.

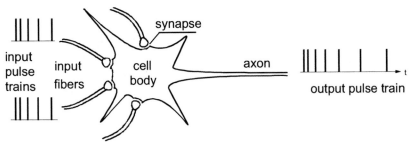

Fig.3.1. The outline of the biological neuron

A simplified view of a biological neuron shows a cell body with one output fiber, the axon, and a large number of inputs that connect to the cell body via special contact points, synapses. The input signals as well as the output signal are in the form of constant amplitude electrical pulse trains with variable pulse repetition rate. All right then, the neuron communicates with electrical pulses, but if this is the case, where is the return wire? There is only one output line — the axon — and every electrical engineer knows that electric current can only flow in closed circuits.

The neural fibers like the axon are not simple conductors. They have very high resistance and the electric pulses propagate as local potential differences between the outer surface and the core of the fiber, therefore no return wire is needed. Due to this propagation method the physical speed of these pulses is limited, about 1 to 100 meters per second depending on the type of the fiber.

The input contact points, synapses, are small caps between the input fiber and the actual neuron. There are two kinds of synapses, namely electrical ones and the more common chemical ones. Electrical synapses pass the electric signal via a low resistance channel while chemical synapses rely on molecule transmission. Chemical synapses introduce a millisecond delay to the propagating signal while electrical synapses do not have such a delay. An electrical input excitation at the chemical synapse leads to the release of certain chemical molecules, neurotransmitters, that bind to the neuron body side of the synapse causing there neural excitation or inhibition depending on the type of the synapse[74]. The beauty of the chemical synaptic operation is that the synaptic threshold for large groups of neurons can be modulated simultaneously by various chemicals[75]. On the other hand individual synaptic transmission strengths can also be modified by neural learning, via the so-called long-term potentiation (LTP) effect, this is the basis of memory.

It seems that the neuron itself is a decision maker, it decides whether to fire a burst of pulses to the output fiber or not to fire. This decision depends on the excitation level of the neuron. This in turn depends on the number of excitatory and inhibitory input signals and their repetition rates as well as on the learned synaptic strengths and chemically modulated general synaptic threshold levels. In this way the neuron adds all the effects together and fires if the resulting excitation level is higher than the given threshold at any moment. The resulting output pulse repetition rate is proportional to the excitation level.

There are about 100 000 000 000 neurons in the human brain. Each neuron is connected via some 1000 to 10 000 synapses to other neurons and thus the human brain forms a very large neural network that processes information by coordinated neural firings. The brain is not a homogeneous unit though. There are distinct parts and structures like the cerebral cortex, cerebellum, hippocampus, thalamus, midbrain and the brain stem. However, the early pioneers of artificial neural networks did not much consider the modular structure of the brain, instead the neuron and its immediate connections were taken as a sufficient model.

The Artificial Neuron

An artificial neuron is devised to mimic the operation of a biological neuron. Accordingly a typical artificial neuron has a number of inputs each with its own weight value ("synaptic weight"), a summing junction and a threshold or nonlinear output function. Instead of pulse trains continuous signals with variable intensities or values are usually used. The value of each input signal is multiplied by its related weight value and the results are summed together. This sum value is taken to the threshold circuit and the neuron will provide output if the threshold is exceeded. This kind of an artificial neuron was originally proposed by Warren McCulloch (M.D.) and Walter Pitts already in 1943[56].

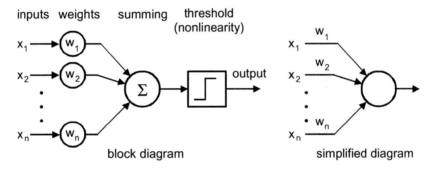

Fig. 3.2. The block diagram of an artificial neuron and its simplified diagram

In fig. 3.2 the inputs x_i and weights w_i have numeric values and the neuron itself computes a numeric output from these.

Artificial neurons like this can be used to recognize and classify objects. Let's consider a simplified example. Suppose that we have cherries and red apples among some other fruits and we would like to have a machine that could sort out cherries and apples by their appearance into their own boxes. For this purpose we need a two-neuron network like the one in fig. 3.3.

In fig. 3.3. we may have a simple image sensor that is able to tell us the color, size and shape of the objects. Let's suppose that each property is indicated by a signal that has the value "1" whenever the property is present and "0" at other times. (In practical applications continuous signal values are often used instead of one and zero.) For instance if the size "small" is detected, then the "small" signal will have the value "1" and the "large" signal will have the value "0".

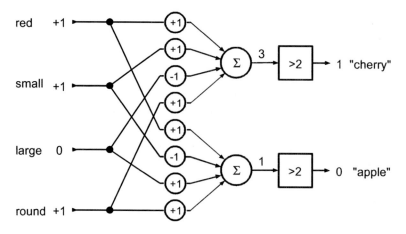

Fig. 3.3. Two-neuron one layer neural network to sort out cherries and apples

A cherry is red, small, and round. Thus /red/=1, /small/=1 and /round/=1. Al the other input values will be zero. The cherry-detecting upper neuron in the figure 3.3 will multiply each input by the related weight value and sum the results together in the following way:

/red/ 1 x 1 + /small/ 1 x 1 + /large/ 0 x (–1) + /round/ 1 x 1 = 3

The sum 3 is larger than the output threshold of the neuron and therefore the output will be 1, a cherry is detected. At the same time the apple-detecting lower neuron will do the computation:

/red/ 1 x 1 + /small/ 1 x (-1) + /large/ 0 x 1 + /round/ 1 x 1 = 1

This sum is smaller than the output threshold and therefore the output will be 0, an apple is not detected.

In the same way we could have more neurons and more input properties to sort out every fruit or other item as needed. The universal recognition problem is thus solved or is it?

The Exclusive-OR Problem

Unfortunately things are not that simple. Consider a case where an item has the property A or B but not both of them at the same time so that when A or B is present, the item is detected but if A and B are present together, then some other item is present. What kind of weight values should we have in order to correctly classify the item? It turns out that no combination of the weight values can do the job, if the property A alone is able to fire the neuron and B likewise, then A and B together will necessarily fire the neuron, too. This is the notorious "exclusive-OR" problem.

What can we do? We could try to detect the special case where A and B are both present and use this detection to force the output to zero. This can be achieved by adding another neuron layer, the "hidden layer", between the input layer (input points) and the output neurons, like in fig. 3.4.

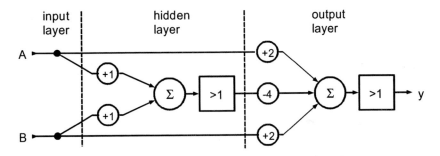

Fig. 3.4. Hidden layer solves the exclusive-OR problem

In this example the hidden layer and the output layer consist of single neurons. It is assumed here that the inputs A and B may have values 0 (input not present) or 1 (input present). The hidden layer neuron detects the case where the inputs A and B are both present by computing the sum of A and B. If this sum is 2, then both A and B are present and the hidden layer neuron output will be 1. Now in this case the output layer neuron will compute the sum

$$/A/\ 1\text{x}2 + 1\text{x}(-4) + /B/\ 1\text{x}2 = 0$$

This sum is below the output threshold and the output will be zero as desired. On the other hand if A is 1 and B is 0 then the hidden layer output will be zero and the output layer sum will be 2, which is above the output threshold. The output will therefore be 1 as desired. Likewise if A is 0 and B is 1 then the output will be 1. It can be seen that the exclusive-OR problem is now solved by the addition of a hidden layer neuron.

Similarly hidden layers with large number of neurons can be used to detect and resolve other unwanted cases, more general cases of the exclusive-OR problem.

Neural State Machines, Recurrent Neural Networks

Feed-forward neural networks like those described earlier are able to recognize and classify static patterns that represent a snapshot in time. This operation is not different from look-up tables and therefore a fully trained feed-forward neural network can be regarded as a look-up table. Thus obviously the computational limitations of look-up tables apply also here.

State machines are more powerful than look-up tables and they can handle temporal sequences. A state machine is characterized by the feedback signal, the state value that is routed from the output to the input.

By using feedback in a similar way the output of a neural network can be made to depend on instantaneous input signals and the past output of the network. This can be achieved by routing the output or part of it via a delay line to some of the input points as in the fig. 3.5. This kind of a system is called a recurrent neural network.

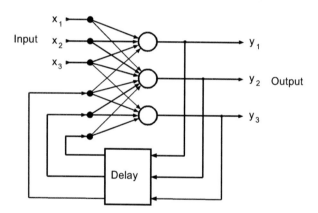

Fig. 3.5. A recurrent neural network. Previous output is routed to input

A recurrent neural network operates as a state machine if the neural weights are fixed and no modification of these takes place during the operation. In this case the neural network constitutes a set of look-up tables, nothing more. The computational limitations of state machines apply also here.

However, it has been shown that the recurrent neural network will equal the Turing Machine if the neurons are made to operate as embedded read/write memory. But just like the Turing Machine a recurrent neural network may not be the most practical (or practical at all) solution for general information processing.

Learning in Artificial Neural Networks

In practical applications we would have a large number of neurons arranged into several layers. Each neuron would also have a large number of inputs and input weight values. As we have seen, the weight values determine the operation of the network, what the network is able to recognize or classify. The neurons and networks seem to be simple enough, but how do we determine the correct values for the weights? This is a real problem. Remember, neural networks were supposed to provide us a road to machine intelligence via self-learning, without programmed algorithms. So, we have to make the networks learn by examples.

Basically there are two ways to make this happen, the supervised and unsupervised method. In supervised training we supply an input and a corresponding desired output to the network. Then the actual output is compared to the desired output and the difference is used to adjust all the weights against each other so that the desired output is achieved. This process is repeated with a very large number of inputs and desired outputs and if all goes well then eventually the weights within the network will converge to values that allow correct recognition or classification of totally new inputs. Alas, an algorithm is needed to adjust the weights during learning. The "backpropagation algorithm"[45] is widely used and perhaps the best-known one.

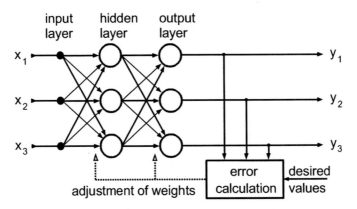

Fig 3.6. The supervised training of the network can be done by calculating the differences between the actual and desired outputs and minimizing them by weight adjustment.

In unsupervised learning no desired outputs are presented. Instead a weight-adjusting algorithm is used that will eventually cause similar inputs to have same output patterns. However prior to training there is no way to tell what these output patterns for each class of inputs will be. Usually this is not a problem though. Again, a very large set of example inputs is needed.

Now there is nothing in the original neuron model of fig. 3.2. that as such could implement any of these weight adjusting algorithms. In fact, weight adjusting is something external to these neurons and best realized by a computer! Therefore in many cases neural networks are implemented as computer programs, as a style of programming and calculation. Somehow this kind of flattens the original idea of non-program computing.

Arguments against Neural Networks

Neural networks can be criticized in this context that they do not really model human cognition. The concept that biological neurons would rely on complicated weight adjusting algorithms can be seriously doubted.

The human brain has some 100 000 000 000 neurons and each neuron may have tens of thousands input synapses and hence input weights. What kind of an algorithm would find a suitable balance between these weights so that the output error would be minimized? Where in the brain would be the computing process that would execute weight adjusting algorithms? The evolutionary feasibility of this kind of algorithm can be seriously doubted. Complicated algorithms do not evolve via trial and error either. Moreover, humans are able to learn with a few examples only instead of training sessions with thousands or hundreds of thousands of examples.

Neural networks are usually used as classifiers and classification-based pattern recognizers. However, classification alone does not constitute cognition. We do not perceive our environment as a collection of labeled objects, in fact we do not even think of the labels — names for the objects — unless we are specially asked for them. Instead we perceive the possibilities of the environment, how to move around, what objects to use and how, what is relevant to the present situation and our needs. We need to process relationships and significance. But, conventional artificial neural networks are not good at processing structural connections.

Conventional artificial neural networks are not very suitable for symbolic processing. The input signals are treated as numeric values and the network performs calculations on these. The input values will not be treated as symbols that have attached rules to be executed. In this way computational neural networks only map arrays of input numbers into arrays of output numbers, this is hardly my idea of thinking and cognition.

Nowadays these conventional artificial neural networks are usually realized as computer programs. Therefore all that has been said against computers as thinking machines should also apply here. In that context neural networks can be seen only as a style of numerical computing and not even really different from statistical computing methods.

This is not to say that symbolic processing with artificial neural networks in the strict computing sense or in the human sense is impossible even though it is not usually done today. A different style of realization would be needed and later on in this book I will outline artificial neural networks that process information symbolically, in the human sense.

PART II

COGNITION AND CONSCIOUSNESS

CHAPTER 4

COGNITION AND PERCEPTION

The Real Processes of the Mind

Cognition is a general name for a collection of processes that are involved in perceiving and thinking. A cognitive system, be it human, animal or robot, utilizes sensors to acquire information about its environment and its own physical status and needs. Multiple sensory modalities may exist; sensors for vision, audition, touch, olfaction, etc. The information from these multiple sources must then be processed so that a unified non-contradicting interpretation of the moment-to-moment situation can be had. This information is further processed so that proper responses can be generated. Cognition answers the questions like "what is where", "what is happening", "what should be done". It utilizes various processes and mechanisms so that most relevant information can be recognized, experience can be accumulated and utilized, deduction and reasoning can be applied. Cognitive processes are studied by cognitive psychology and cognitive neuroscience.

It should be noted that cognitive science is not yet an exact science. The concepts of cognition are based on empirical research but may still be subject to interpretation. Even the basic concepts may have different contents in different textbooks. Therefore the following discussion should be taken not as a statement of exact facts but as a review of ideas and viewpoints, others' and mine, with merits and shortcomings that may remain to be verified by further research and in this case especially by their applicability for artificial machine realization.

The main elements involved in cognition are depicted in fig. 4.1. Information from physical world is acquired by various sensors and this information meshes to the actual cognitive processes. Perception processes interpret the sensory information, selected by sensory attention. Learning and memory processes enable the accumulation and utilization of experience.

Deduction and reasoning allow access to information that is not directly perceived. Judgement processes detect contradictions and evaluate significance. Emotions are here included in cognition for reasons that are explained in chapter 6. Language is a tool for communication, but it is also a cognitive tool. The relationship of consciousness to cognition will be discussed in chapter 8.

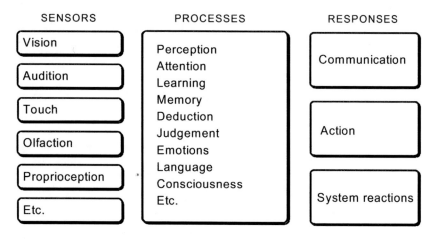

Fig. 4.1. Elements involved in cognition

What is thinking then? Introspection would seem to reveal processes like inner speech, inner imagery, feelings and sensations. Are these also the actual processes of cognition? Is thinking indeed performed by using the perceived entities of inner speech and imagery or would thinking be based on some subconscious underlying processes instead?

There are two major symbolic theories of thought; the natural language theory and the "mentalese" theory. The natural language of thought theory states that thinking is indeed rule-based manipulation of symbols and these symbols are the words of the natural language that we use in our inner speech. On the other hand the "mentalese" theory[33] of thinking states that thinking is rule-based manipulation of such symbols that constitute the vocabulary of subconscious innate language, the "mentalese". Everybody is supposed to be born with this language and its grammar. The natural language inner speech that we have would only be a translation of the innate "mentalese".

In the following I will not accept either of these theories as such, instead I will approach the question of language of thought and communication from a unified point of view that can also be implemented artificially.

From Sensation to Perception

Our senses provide us a continuous flow of sensations. Our eyes deliver us visual information worth millions of picture elements every second. Our ears deliver us streams of sounds. All our senses compete for our attention each with their own signals.

When we look around we seem to perceive and recognize effortlessly the things around us. We recognize also sounds, be it wind, a motor car, airplane, a cricket or someone talking to us. Our senses provide us a moment-to-moment picture about the world around us; a picture where everything has a meaning immediately available.

Perception seems to be such a simple and effortless process. However, our eyes do not tell us what objects they see, any more than a camera would be able to tell what objects it is capturing. Our ears do not tell us who is saying what, any more than a microphone could do. Our senses provide us only the raw or somewhat preprocessed data and it remains for further processes to interpret what is hidden in this information. Therefore perception processes are needed to make sense of the streams of raw sensory information.

For instance a sensed image consists of a large number of picture elements each having its own light intensity value. These perhaps noisy picture element values must now be processed to indicate what is the background and what and where are the objects.

Object recognition is further complicated by the varying appearance of objects in a given category. Also the viewing angle and distance may change so that the sensed image of a given object will never stay exactly the same. The situation is rather similar with other sensory modalities, too. Therefore the cognitive system must somehow be able to find sensory patterns common to each category so that objects and entities could be recognized even when the sensory information varies and is contaminated by noise. So far this would seem to be a demanding but nevertheless straightforward pattern recognition task.

Fig. 4.2. Objects of a category may vary in appearance

Technical pattern recognition involves the exact recognition of an object. A vending machine that accepts bank notes must recognize the notes without error, it must not accept fakes, pieces of paper that are made to look like the real notes. Sometimes it does happen that a vending machine rejects a real bank note, one that we humans can easily recognize as a genuine bill. In many cases human perception does indeed a better recognition job. Is human perception really based on a perfect errorless pattern recognition process? Consider the following picture, what do you see?

Fig. 4.3. Are these faces?

There are three faces in this figure, right? However, if I were to design a bank note pattern recognizer that would recognize these as faces then most probably my boss would not be very delighted. Indeed, a pattern recognizer that makes a face out of a circle and few lines is not a good one because it could see faces everywhere. Most probably it would also accept almost everything as a bank note. But, we can see that these are faces, we can even tell if they are happy or sad. On the other hand we also know that these are not real faces, they just remind us of ones. So what is going on?

First, it seems that our visual perception is based on elementary feature detection and in this case some of the basic facial features are reproduced. Secondly, our pattern recognition process is not perfect, as only few cues are enough for the perception of an object. Thus, *a percept is not an absolutely recognized real thing, it only reminds us of those.*

Now consider this:

100 B00KS

It says: "one hundred books", doesn't it? However, the zeroes in "100" have exactly the same appearance as the "oo:s" in "books". Yet we do not have any difficulty to recognize the first "oo:s" as zeroes and the second "oo:s" as letters even though there is nothing in their appearance to suggest one interpretation or the other. In this case, however, the correct interpretation comes instantly from the context and even without our awareness of the process. Similar examples can be found in spoken language. Similar sounding words can have different meanings depending on context.

Thus we can see that straightforward pattern recognition, no matter how perfectly implemented, is unable to cope with situations where the recognized object has multiple interpretations depending on context. In this case the patterns just do not contain the required information for correct interpretation. Extra information must therefore be extracted from the context or elsewhere.

Which letters are hidden under the black square?

We would not be surprised if the letters "*ION*" were revealed. If the black square was to evaporate gradually and we would start to see some letter-like patterns appearing, we would be inclined to perceive especially the letters "*ION*". Perception utilizes prediction; expectations will affect what will be perceived.

Also attention affects perception. This can be illustrated by the following quite familiar images, the face/vase and the bird/rabbit picture.

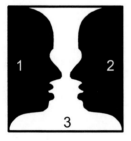

Fig. 4.4. Left: Two faces or a vase? Right: Is it a bird or a rabbit?

Two faces or a vase? The mental interpretation can be changed by shifting attention between points 1, 2 and 3. The bird/rabbit picture has also two interpretations. If attention is focused to the point *a* then a bird is usually perceived. If attention is focused to the point *b* then a rabbit may be perceived.

So, perception is not straightforward pattern recognition, instead it is a process that combines the effect of sensory information and the system's inner information. A percept is not a representation for an absolutely recognized entity, it is only an impression that may change according to context and other cues. The perception process does not only utilize sensory information, it also utilizes the effects of context, expectation and attention.

We can now sketch a system that does this, fig. 4.5. This system consists of sensors, sensory signal preprocessing and an actual perception process unit that combines the effects of sensory information and the system's own information.

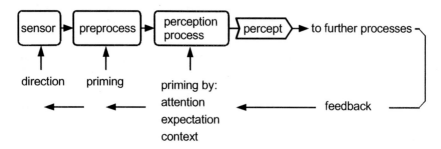

Fig. 4.5. Sensory and perception processes

In the system of fig. 4.5 raw sensory information is derived from sensors. This information is preprocessed by appropriate processes like filtering, transformations and feature extraction. The actual perception process combines the preprocessed sensory information and the system's inner information. This process can be additive; the sensory signals and the system's feedback signals are added together so that the common part of the sensory and feedback information is amplified, "primed". This addition may be followed by threshold operation so that the primed part will constitute the actual product of perception, which is called the percept. The priming by feedback can be extended to the preprocess where it would adjust the preprocessing parameters. Priming could also determine the direction of the sensors where applicable, for instance the direction of the head and eyes.

The priming will, so to say, help the perception process to find the relevant part of the total sensory information. It is like a teacher giving cues to an unsure pupil. The feedback reflects the system's expectations and therefore what we get now is a percept that should be consistent with the cognitive state of the system. This percept should now fit the puzzle provided by the sensory modalities, context and system's accumulated knowledge.

The priming by expectation has indeed such a power that it may even override the actual sensed information. The difficulty of proof-reading is one example. Misspelled words are hard to detect because we know and expect the correct words and therefore we will also perceive them as such.

I would even say that we do not recognize objects as such at all. They just remind us of something and thereafter we can recognize them because we know what they are. This knowledge will help us even when the object is sensed only partly, priming will provide the missing information. Most striking example of this would be the mime theater; we will recognize objects that do not exist at all.

A plausible perception mechanism must also be able to explain the phenomena of illusion, hallucination and dreams. An illusion arises when sensed signals are interpreted to depict something that does not in fact exist at that moment. A hallucination is the perceived presence of something that does not exist, in this case nothing is actually sensed. The perceived entity is totally a product of imagination, however this is not recognized and the imagined entity is interpreted as an actually sensed external entity. Illusions and hallucinations may be experienced in the awake state. Any normal person may have illusions, but hallucinations are only experienced during altered mental states, due to exhaustion, disease or drugs. Dreams are hallucination-like perceptions that are experienced during sleep.

It can be directly seen that illusions are possible in the proposed perception mechanism. In fact every percept is a kind of illusion, created by the priming process. Normally this illusion is in good agreement with the facts, but sometimes an alternative, false interpretation can be sustained and we perceive something that actually does not take place.

The proposed perception mechanism is also able to perceive the inner states of the system. This takes place when there is no external input or the external input is somehow suppressed. In that case the percept is determined solely by the feedback, it would be a product of imagination. Because of this possibility additional signals that indicate the actual origin of the percept must be assumed so that the distinction between externally evoked and internally evoked percepts can be established. The activity level of the actual sensor and preprocessor could be one indication that something is actually sensed. Now a hallucination would arise if this distinction between external and internal evocation were somehow to fail.

Technically dreams are hallucinations that we have during sleep and their perception can therefore be also explained by this same mechanism. A model for the higher aspects of neurocognition would be needed for the explanation of the various other aspects of dreaming.

Sensory Match, Mismatch and Novelty Detection

Is this a book? How do we know what to answer when this question is put to us and an object is presented? Likewise, when we are searching for something, how do we know that we have found it? Obviously we may have an inner idea what the object in question should look like. We may compare this idea to the percept of the presented object and deduce the answer. But how do we do this deduction? We may try to accomplish this task in any way we can, but at the end we must have a basic inner process that is able to tell directly whether the percept and the inner idea match, even partly. Luckily nature has provided us a way to do these comparisons easily and usually without any conscious effort, we just know whether a perceived object is or is not the one in question. However, we are not able to perceive how this mechanism works. These facts would seem to point towards a low-level process, one that we do not have to learn consciously. Indeed, this kind of mechanism can be realized within the framework of the perception process model as a very basic hard-wired process. In the perception process model the inner idea would manifest itself as the feedback signal, thus the match detection would reduce to a comparison between the sensory signal patterns and the feedback signal patterns.

We can now define match, mismatch and novelty conditions that correspond to the sensory/inner signal relationships as follows:

sensory signals	=	inner signals	⇒	match
sensory signals	≠	inner signals	⇒	mismatch
sensory signals	–	no feedback	⇒	novelty

These sensory match, mismatch and novelty signals can be derived from the feedback reentry point where the sensory and inner signals meet as depicted in fig. 4.6.

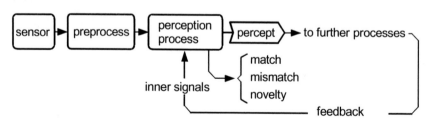

Fig. 4.6. Derivation of sensory match, mismatch and novelty signals

There is empirical evidence on neural match/mismatch functions in the brain. Actual match/mismatch/novelty related EEG signals have been recorded by Risto Näätänen and others. Näätänen has recorded a so-called processing negativity signal (PN) that seems to be related to match-process. He has also identified a so-called mismatch negativity signal (MMN), a negative component of a cortical event-related potential that can be recorded when any discriminable change occurs in a sequence of repetitive homogeneous sounds[81,82]. This signal does not depend on conscious attention.

Later on it will be proposed that the match/mismatch/novelty signals are also related to attention control. It will also be proposed that match-condition were accompanied by slight pleasure (match-pleasure) and mismatch-condition were accompanied by slight displeasure (mismatch-displeasure).

Visual Perception

Eyes allow us to determine visually what is where and how it is moving. Incidentally this is the same information that is collected by a modern radar. Objects are represented by their size, shape, color and texture, their location is indicated by direction and distance and their motion is indicated by apparent speed and direction.

The eye can be compared to a camera or, better yet, to a modern semiconductor video camera. The image of the viewed object is projected by the lens on the retina, a layer of light sensitive receptors; rods for low illumination grey scale detection and cones for daylight color reception. The image on the retina is upside down, but this is of no consequence for the brain and no image reversal is needed. The resolution of the eye depends on the density of the receptors on the retina; the more receptors per square millimeter, the sharper picture. The highest resolution area, the fovea, is located in the center of the retina.

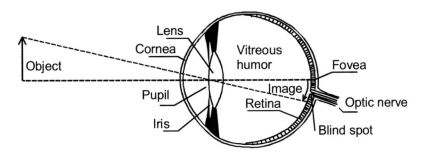

Fig. 4.7. The structure of the eye

Here we must also notice an unfortunate "design error" of the human eye. The light sensitive receptors are not facing the lens, instead they are upside down so that the preprocessing neurons and wiring face the lens and obstruct the very light that should be sensed. The wiring is collected and combined into the optic nerve at one point on the retina, not very far from the fovea. At this point where the nerve leaves and goes through the retina there can be no light sensitive receptors and therefore this point is a blind spot. If we had better resolution and thus more light sensitive receptors around the fovea as well as at the periphery then there would be more fibers obstructing the fovea and the blind spot would be larger, too. Thus the good resolution of the fovea would be compromised. Lower resolution at the periphery means therefore that the resolution of the fovea is not compromised, given the unfortunate overall design. But it also means that less visual information is to be processed and thus also less neurons are needed, leading to improved biological economy.

There are some 120 million receptors all together on the retina but after some initial processing only about one million optic nerve fibers leave the eye[36]. We can think these million signals carried by these fibers as the preprocessed picture elements that are forwarded to the brain for further processing.

Visual perception process uses the retinal image to determine what is where. First the objects must be separated from the background and then the objects must be recognized. This is not a straightforward task. The recognition of an object calls for the proper binding of the individual picture elements that depict the object. The task is not made any easier by the fact that objects are not always seen from the same viewing distance and direction. Therefore the apparent size, position and angle of the object varies on the retina. The illumination may change, sometimes only noisy images can be seen. Objects may also be obstructed by others. The identity of the object must be preserved; an object moving from point A to point B must be recognized as the same. Yet the perception process must comply with these challenges.

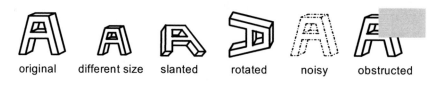

original   different size   slanted   rotated   noisy   obstructed

Fig. 4.8. The challenges of visual perception due to retinal image variation

Complete scene understanding calls for the recognition of the individual objects and their relations as well as their associated meanings, significance, past and possible future. This task is considered here as a cognitive process, one step higher from the basic perception and will be discussed later.

Another problem with visual perception is the large amount of data. If the images provided by the eyes were to be stored and used as such, at a rate of, say, one picture per second then the data rate would be at least 1 million bits per second, probably even greater. Therefore during a day about 60–100 gigabits would have to be stored. The human brain has about one hundred million million synapses and if each synapse were to store one bit then the storage capacity of the brain would be 100 000 Gigabits. From these numbers we can conclude that if the brain were to store all images as such, then the brain of an infant would be completely full after three years. True, this is a rough estimate, but if I were certain that the brain does indeed store images as such then I would strongly advice anybody not to watch those television soap operas night after night, as this would lead brains to overflow in no time at all. However, everyday experience shows that many of us (especially those who do watch soap) seem to be empty-headed most of the time. Therefore it would seem to be clear that the brain does not store imagery in this way; the brain is not a video recorder and images are not stored as bitmaps. Instead visual objects and their relationships are derived and finally the gist of the story may be stored for a while.

How are objects recognized? It is known that in the visual cortex, especially that corresponding to the fovea, there are neuron groups that recognize small visual features like short lines and their orientations. It is assumed that the object recognition is a hierarchic process starting from elementary features and basic component forms. Simple detectors are sufficient for elementary feature detection and yet it is possible to construct a large number of basic component forms with a limited number of these elementary features. Then again these component forms may be combined in different ways to depict real world objects. In this way the recognition problem is reduced to the recognition of the limited number or elementary components.

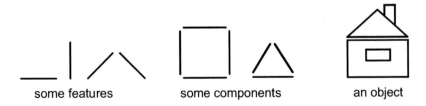

some features   some components   an object

Fig. 4.9. From detected features to basic component forms and real objects

Is the component form phase necessary in the process? Wouldn't it be possible to recognize objects directly as combinations of elementary features? Small objects may indeed be recognized in this way. However, due to the small size of the fovea it may not be possible to perceive all the features of large objects at the same time, in that case the only way to succeed would be to scan individual parts of the object and see how these are connected.

The visual world around us is not exactly a line drawing even though contours that separate objects from background may be easily detected and described by elementary features. The visual world consists of areas of more or less constant shading, color and texture, too. Therefore binding of adjacent pixels with similar grayscale and color values will help the overall recognition process.

Apparent size variation and rotation may be compensated by cortical mappings that separate shape, size and angle information.

The detection of distance is based mainly on cues from perspective, on the covering and uncovering of background objects by foreground ones when we turn our head, stereoscopic vision, the angle of the eyeballs and the line of sight.

Things that are near us appear larger than those further away. Also when we move nearby objects seem to go by faster than distant ones. These are manifestations of perspective.

When we are moving or turn our head the foreground and background objects will be in apparent motion in relation to each other. The background objects will be covered or uncovered by the foreground objects. Obviously the covering objects are then nearer to us than the covered ones. This process works also with one eye only. This same principle works also when the objects themselves are moving covering and uncovering each other.

Sometimes it would be better if the distance of various objects could be estimated without movement. It is not necessary to move the head if you can get different angles of view by other means. Stereoscopic vision does this with two eyes that provide slightly different images. The differences between these images, so-called binocular disparities, allow us to see a little bit behind foreground objects and in this way determine which objects are closest to us.

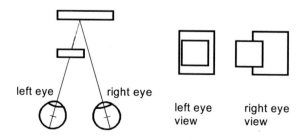

Fig. 4.10. Stereoscopic vision by two images

When we look at distant objects the lines of sight of our eyes are practically parallel. When we look at objects that are nearer, we must turn our eyeballs so that the object will be focused at the fovea of each eye. This angle of the eyeballs can be sensed by the tension of the respective muscles and will give an indication of the distance of the object. Obviously the accuracy of this method is the better the nearer the object is as the angle of deviation gets larger.

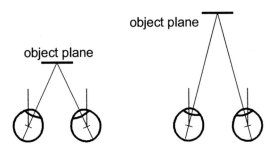

Fig. 4.11. Distance estimation by the line of sight angles

In many cases to look closer means to look down. For instance if we have things on the table we must look more downwards the closer these things are to us. This correlates well with the possible trajectory that we should use the reach out for that thing with our hand. The gaze direction especially for nearby objects is connected to the motor commands that enable the grasping of the object.

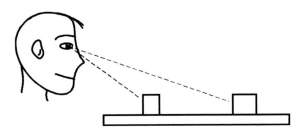

Fig. 4.12. We must look downwards for the objects near us.

The perspective gives additional cues about the distance of objects. The apparent size of known objects is inversely proportional to their distance, the further away they are the smaller image is projected on the retina and the smaller they look.

The detection of motion of visual objects is based on three principles. Horizontal and vertical motion of objects may be detected by the static eye; specific cells respond to the motion of the object image on the retina. The same motion can be also detected by tracking the object by turning eyes and head so that the image of the moving object remains stationary on the fovea. The motion in now determined by the required rate of the motion of the eyes and the head. When an object is moving directly towards or away from us neither of these principles is so efficient. In this case the motion can be determined by the change of the apparent size of the moving object on the retina.

Visual change detection ability is essential for the guidance of visual attention and obviously would be useful for other purposes, too. Would this ability be a low-level process or based on the comparison of stored visual imagery? This can be found out experimentally. However, experiments that can show this have produced quite surprising results.

Simons and Levin devised an experiment where an unwary subject was confronted by a stranger asking for directions. While the subject was talking to the stranger two men carrying a door passed by and covered the stranger for a short while. This act was used to conceal the substitution of the stranger by another quite different looking person. About 50% of the test subjects failed to recognize the substitution of the stranger and went on to give instructions to the new stranger as if nothing had happened[92]. So, what is wrong here, why did the subjects fail to notice a major change in the situation?

In other experiments it has been found out that it is indeed very difficult to notice a change in a scene if the scene is blanked out for a short while before the change. This effect is called "change blindness".

Fig. 4.13 depicts one kind of experiment that demonstrates change blindness. In this experiment a sequence of images is displayed on a computer screen. If a blank image is displayed immediately before a change in the picture then the change will be hard to notice.

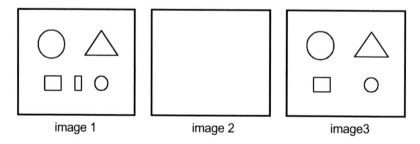

image 1   image 2   image3

Fig. 4.13. Change blindness. It is very difficult to notice the difference between image 1 and image 3 when the images 1, 2, 3 are sequentially presented on a computer screen.

However, if in the experiment of fig. 4.13 the blank image is left out and the image 3 is displayed immediately after the image 1 then the changes will be noticed readily, fig. 4.14.

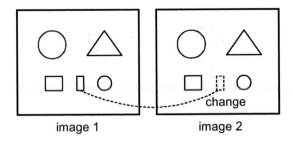

image 1   image 2

Fig. 4.14. Changes in continuous scenery will be noticed readily

It is obvious that change blindness is caused by scene blanking, even of short duration. The simple conclusion is that visual change is detected directly at the picture element level by comparing the present picture element value to its delayed version. Larger area changes can be detected by combining these picture element value changes. All this can be accomplished without the need for any actual image memories. Thus in the example of fig. 4.13 the change between the images 1 and 3 are not detected as the change detection process indicates complete change between images 1 and 2 and another complete change between images 2 and 3. Therefore it is quite clear that no indication about the difference between images 1 and 3 can be had by this mechanism.

These experiments also show that visual scenery is not stored in the brain as sequences of actual imagery, instead only general ideas about what is where may be stored. The environment itself serves as the supply of fine details and these are inspected and memorized only as needed. We can describe only with limited accuracy what we saw before a couple of scene changes. In fact, photographic storage of imagery would be detrimental to our cognitive capacity due to the immense need for memory space.

Gun sights have crosshairs to facilitate the exact pinpointing of the target. Humans need to pinpoint objects as well, pick up visually just the instantaneously relevant object and exclude others for that moment. The human eye does not have crosshairs, so what property of the eye is used for that purpose?

The human eye has a very wide field of vision when compared to that of most cameras. However, we do not see everything in our view sharply, only the center part of the visual field is seen with full resolution. This angle of sharp vision is surprisingly small, a fact that easily escapes us as we scan our surroundings subconsciously and get the illusion of full field full resolution perception. The fact that the fovea or the high-resolution area of the eye is indeed very small can be easily verified by a simple test. The reader is asked to gaze directly to the center X in the figure 4.15.

K	Q	E	K	P	W	L	K	J	G	F	H	S
O	W	E	G	J	L	K	C	V	I	P	Z	Q
I	Q	Q	R	W	N	P	Z	O	M	S	C	H
D	V	I	O	T	J	U	A	P	Q	A	E	D
M	A	Q	R	K	N	A	L	F	P	U	R	K
B	D	W	U	G	M	X	K	B	L	G	I	O
C	S	U	H	D	R	E	S	Q	E	H	U	M
A	F	K	T	Z	V	C	T	V	B	P	T	P
N	B	Q	H	O	K	F	S	A	T	D	O	A
Z	G	W	E	J	R	U	I	P	N	M	H	N
G	S	B	V	C	Z	G	R	S	E	F	L	T

Fig. 4.15. The angle of sharp vision is surprisingly small. How many letters around the center X are you able to recognize without moving your eyes?

How many letters around the center X are you able to recognize without moving your eyes? Not many. The point is that in order to see an object accurately we have to direct our gaze precisely on it.

Thus the limited size fovea doubles as the crosshair allowing us to pinpoint objects accurately. It also excludes the accurate perception of objects in the periphery. This property of the eye has enabled a whole industry — we can enjoy comic strips! Indeed, there would not be much point in comic strips if we were able to see and perceive the whole page at one glance. The artists would have to hide the gag behind the page, not very practical.

The fact that we have to direct our gaze precisely does have one drawback. Our direction of gaze gives away what is being looked at. On the other hand this also allows pointing, we can direct the visual attention of others by pointing at the intended object.

Visual attention can also be directed by peripheral vision. Even though the resolution of the peripheral retina is low it is still very sensitive to change, moving objects may be detected even though they cannot be recognized very well. Motion, sudden bright light, visual change, bright patterns, etc., detected by peripheral vision can cause the redirection of visual attention.

Visual attention can also be directed by sounds. The ears are able to determine the direction of a sound and likewise a sudden sound makes us to turn our heads and gaze towards the supposed sound source.

Also internal factors guide visual attention. We may be executing a task that necessitates continuous visual attention redirection.

Some tasks like reading call for fixed pattern visual attention redirection while others like scene understanding may require flexible redirection.

We focus visual attention on objects that are relevant to our needs or are emotionally significant; novel, pleasing, beautiful, sexy, strange, threatening, etc.

Auditory Perception

The sound that enters our ears is a complicated time-varying signal consisting of the sum of all instantaneous sound waves around us. The task for the auditory perception process is to separate the relevant sounds from this sum signal and locate their source. The complexity of this task may be better appreciated by studying a picture of the temporally varying sum of sounds.

Fig. 4.16 depicts a short sample of a possible variation of air pressure intensity due to the combined effect of separate sounds. This is like the sound signal at our ears, yet this is not what we hear. We can get an idea of the true nature of sound signals by placing our finger on a loudspeaker membrane and feeling the vibrations. These vibrations are similar to those at the eardrums, yet no perception of separate sounds will arise; the mere sensing of vibrations will not suffice. Still, our ears are able to segregate separate sounds and provide us with their directions even though at each moment only the instantaneous sum of these sounds are received.

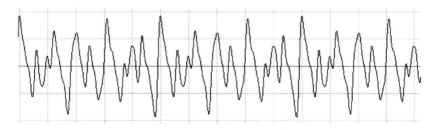

Fig. 4.16. The challenge of auditory perception. What sounds give this combined time varying air pressure signal at the ear?

The old wisdom is: the sum tells little of its components, any large number may be the sum of numerous combinations. The signal received by our eardrums is exactly this kind of a sum, how can we make any sense of it? To answer this question we must consider fundamental properties of sound signals.

The simplest possible signal is the sine wave consisting of a single frequency. If a system is to hear anything then it must be able to resolve solitary sine wave signals.

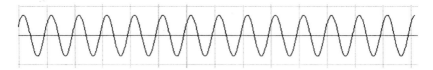

Fig. 4.17. Sine wave signal (500 Hz, 2 ms per division horizontally)

The sine wave can be described completely by two parameters; its frequency and amplitude; the vertical peak-to-peak value. A single audio sine wave sounds like a whistle. The human ear is tuned to distinguish sine waves within the audible frequency range, a kind of frequency analysis is performed. Young people with good hearing (that was before the rock concert era) used to hear sine wave audio signals with the frequency range of about 20 Hz to 20 kHz.

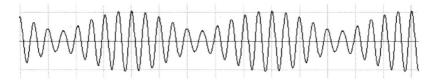

Fig 4.18. The sum of two sine waves (1 kHz + 1.1kHz) with the relative amplitudes of 1 and 0.5

Two sine waves will be heard as separate tones if their frequencies are far apart. However, if the separation of the frequencies is small enough the signals will fuse into a single tone.

It can be seen that the sum sound is periodically strengthening and weakening. The frequency of this effect is the difference of the frequencies of the added signals. This effect, "beating", can even be heard (the "wah-wah" of the sound) as any one who has tuned a guitar knows.

As the separation of the frequencies gets larger, separate tones will eventually be heard but before that the perceived sound gets a rough displeasing quality, the ear cannot decide whether there is a single tone or two separate tones and which of these should get the focus of attention. This effect is also called dissonance.

Pure sine wave signals are rare. Natural sound sources generate periodic sound signals that do not have the clean undistorted appearance of the sine signal. However, these kind of periodic signals can be decomposed into a sum of pure sine signals. These component signals consist of one fundamental sine wave and a number of its harmonics, sine waves with frequencies that are exact multiples of the fundamental frequency. This is not due to some strange whim of nature, this is due to a mathematical property of periodic signals; every periodic signal can be represented as a sum of a fundamental sine wave and its harmonics. As an example the sum of two sine waves is presented here, namely the sum of a fundamental sine wave and its harmonic with a frequency that is twice the fundamental one.

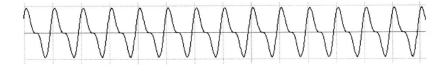

Fig 4.19. The sum of fundamental sine wave and its second harmonic (500 Hz + 1kHz, with relative amplitudes of 1 and 0.5). This is perceived as a single tone, a "bright whistle".

We can see that the sum of a fundamental sine wave and its second harmonic is distorted from the clean sine, therefore the harmonic component is sometimes called harmonic distortion especially if caused by imperfect amplifiers, etc.

Just as two sine waves with the same frequency are fused together, so also a fundamental signal and its harmonics are fused into a perceived single tone with a timbre that is determined by the number and amplitude of the harmonics.

What happens if we change slightly the frequency of a harmonic? Fig. 4.20 depicts the sum of the fundamental frequency of 500 Hz and another frequency 1.1 kHz, a near harmonic. In this case no fusion into a single tone will take place. Two distinct tones will be heard, with lower and higher pitches. However, if the higher frequency deviates less than about 3% of the correct second harmonic value fusion will take place[59].

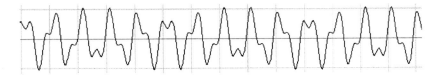

Fig 4.20. The sum of a fundamental sine wave and a near harmonic (500 Hz + 1.1kHz, with relative amplitudes of 1 and 0.5). This is perceived as two separate tones.

Now back to the original problem of fig. 4.16. The human auditory preprocess can extract the components of a sum by performing frequency analysis, by distinguishing the individual frequency components of the sound wave sum signal. In this example the component frequencies and their relative amplitudes would be those given in fig. 4.21.

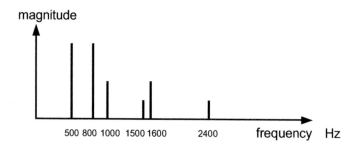

Fig. 4.21. The frequency components of the signal of fig. 4.16.

We can determine that there are two groups of signals here. One group would have 500 Hz as the fundamental frequency and the harmonics 1000 Hz and 1500 Hz. The other group would have 800 Hz as the fundamental frequency and the harmonics would be 1600 Hz and 2400 Hz. The fundamental frequency and its harmonics for each group would be fused as seen before and instead of six separate component signals two distinct tones would be heard. This is a good step towards the problem of sound separation.

Natural sounds are not necessarily continuous whistles, instead they are more like short bursts with varying contour shapes, or envelopes, as they are called in this context.

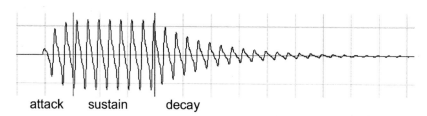

Fig. 4.22. A burst of sound

A burst of sound can be characterized by its total duration, maximum amplitude, the duration of the rising edge or attack, the duration of sustained phase and the duration of the trailing edge or decay. Attack, sustain and decay give a specific coloration to the timbre and provide additional means to distinguish different sounds or musical instruments from each other.

Rhythms are temporal patterns of sequential events and the intervals between them. Auditory rhythm perception is an essential component of cognition; just as visual object recognition relies on the pattern making of parallel visual stimuli, so the rhythm perception process forms patterns in temporal duration, so allowing the categorization of sequential auditory stimuli.

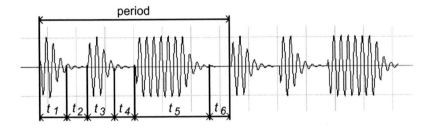

Fig. 4.23. A rhythmic sound pattern

Rhythms are characterized by the relative duration of the constituent events as well as the relative duration of the intervals in between. The constituent events themselves are not essential, be they drumbeats, foot taps, or whatever, the rhythm can be the same.

In the example of the repeating rhythm pattern of fig. 4.23 the rhythm consists of the periodically repeating sequence of sound bursts and their intervals. Each part of the sequence is characterized by its temporal duration, the first sound burst has the duration of t_1, the first interval has the duration of t_2, etc. However, a given rhythm may be played fast or slow, that is, each duration may be divided or multiplied with a constant without any change in the rhythmic pattern. Therefore the absolute lengths of the parts are not relevant, the rhythmic pattern is determined by the relative timing of the sound bursts and intervals. Rhythmic patterns are time-scale invariant.

In practice there are perceptual limits for the tempo of rhythms. Auditory rhythms that are too fast fuse into tones as the temporal resolution of the ear no longer suffices for the detection of individual beats. Auditory rhythms that are too slow disintegrate into discrete events due to the limitations of the auditory short-term memory.

The perception of rhythm involves the detection of the timing of the events and intervals. A repetitive rhythm soon creates expectations of its continuation, which will be matched against the actual sensory input.

The principle of rhythm perception is depicted in fig. 4.24.

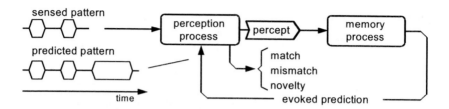

Fig. 4.24. Predictive perception of rhythm

The predictive perception of rhythms creates match, mismatch and novelty states that help focussing of attention and sound stream segregation. Rhythms are also important in speech recognition. This is also manifested in rhymes, last words of lines of verse evoke expectations for the last word in the next line.

The prediction of longer sound patterns calls for perceptual memory. This auditory perception memory is called echoic memory. It allows the recollection of sounds for a limited time with great fidelity. Due to echoic memory it is possible to direct attention retrospectively to what just has been heard. Sometimes we may ask, "what did you say" and at the same time realize that we are able to replay the sensation of what was said and be able to make the correct perception this time.

The direction of the incoming sound can help to locate the source of the sound. Sounds can also be segregated by their direction of arrival. The direction of a sound source is determined by the sound arrival time difference and intensity difference between the ears.

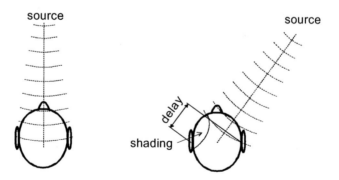

Fig. 4.25. Determination of the direction of sound

If the sound source is directly ahead then there will be no difference between the sounds that reach the ears. However, if the sound is coming from either side then the sound reaching the opposite ear will be delayed in respect to the nearest ear. The head will also shade the sound waves so that the sound reaching the shaded ear will be attenuated. Low sound frequencies are less attenuated by the shading but are sensitive to the delay. On the other hand high frequencies are easily attenuated. It is easy to show that the direction sensing is indeed sensitive to the intensity difference between the ears. This can be simply demonstrated by a stereo system with a balance control. The balance control adjusts the relative intensities of the right and left channel and when balance control is adjusted the apparent position of the sound source seems to move along the line between the loudspeakers even though no delay difference is introduced.

Is the sound coming from front or behind? If we keep our head still the delay and shading mechanisms cannot resolve the front/behind problem. On the other hand our external ears filter sounds slightly differently depending on their direction of arrival. This can give some cues but even so when we keep our head still the front-back direction is difficult to determine. However, as soon as we turn our heads even slightly the case will be resolved by the delay and shading mechanisms.

The distance of a sound source can be estimated by the qualities of the sound, its intensity and coloration. Audio frequencies are attenuated differently over longer distances, therefore sounds with far away sources are not only weaker, they also lack higher frequency content. Multipath propagation from the source to the ears introduces echoes and the relative intensity of these and the direct sound also give distance cues.

Speech recognition is a special case of auditory perception. Here one basic problem is again the sound separation from background sounds, be it other speakers, noise or music. A solution to this problem may be found from the principles outlined above.

The principle of speech recognition seems simple enough if we have a clean single speaker stream of speech. Words consist of phonemes, the "spoken letters". There are only a limited number of these to be recognized. Individual solitary phonemes can be recognized by their typical frequency spectrum and amplitude envelope characteristics. The division between syllables, words and sentences can be detected by their timing, intonation and rhythm. So far so good, but in real speech the phonemes are pronounced in a multitude of ways depending on the adjacent phonemes. Male and female speakers have different voices, people may speak slow or fast, not all phonemes are always pronounced, individual words are pronounced without pauses, the pitch of the speech may change within sentences, etc. This all makes direct speech recognition a complicated if not impossible task. Thus additional cues are needed.

In principle, auditory words are sound patterns that we could produce ourselves. Therefore if a heard sound pattern is able to evoke a motor sequence that corresponds to a word then obviously this cross-connection may be used to help recognition. However this is not exclusive, for instance I cannot pronounce Xhosa or Siswati language words with the click sound, yet I used to be able to recognize a couple of them.

Those who have studied foreign languages have noticed that at first when you hear a foreign language word you may not even be able to tell what phonemes are involved or how to write it; only after you know the word you are able to do this. Also when you are listening to a foreign language conversation, you may find it difficult to understand it unless you know the topic. Examples like these highlight the importance of expectations created by context and background information. Therefore real speech recognition does not rely only on the recognition of phonemes or their combinations, the background information is also important. We recognize words because we know what to expect.

Touch Perception

Touch or haptic perception is based on mechanoreceptors in the skin. A number of different receptors exist with different response times and sensitive areas. These receptors are sensitive to the force that the surface of an object exercises on them while being touched. The fastest receptors, Pacinian corpuscles, are even able to sense vibrations over wide areas of the skin[77]. The density of these receptors in the skin is not uniform, high numbers of receptors can be found especially on the fingertips. The pressure sensation is augmented by temperature sensation. There are separate receptors for low and high temperatures.

The texture of an uneven surface can be sensed by touching it with a fingertip; the pushing of the finger against the surface creates varying miniature forces against the touch sensors according to the ridges and valleys of the surface. These force patterns represent the surface texture.

The sensing of the form of a real three-dimensional object calls for additional geometric position information. This information is available from the proprioception senses, sensors that indicate the position of body parts, here especially hands and fingers, in relation to each other and to their previous positions. The touch perception is now used to indicate the contact to the sensed object as well as the surface texture information at each position. The generated pattern of hand and finger positions will indicate the shape of the sensed object. This process creates a mental "image" of the sensed object and requires thus memory.

Multisensory Integration

The brain integrates percepts from the various sensory modalities into a consistent picture of the world. A heard sound is associated with a visually perceived source, a touch sensation is associated with a visually perceived object, the smell of a rose is associated with the visual and tactile percepts of a flower in your hand. In this way multisensory integration would seem to be a straightforward task of binding; each sensory modality would operate on its own and only the resulting percepts would be bound together.

However, multisensory integration is not that simple. Instead of mere binding a percept from a given sensory modality is able to affect and assist the very perception process of another sensory modality. This assistance helps to resolve any ambiguities especially when the assisted modality is not able to perceive properly on its own, like seeing in darkness or hearing under noisy conditions. Normally this kind of multisensory integration leads to consistent unified perception and we will not even become aware of its workings. Sometimes however, multisensory integration leads to false percepts or even illusions. The auditory sensory system is able to determine the direction of a sudden sound, yet the perceived direction may be that of a visually perceived sudden change that coincides with the sound. The ventriloquist's dummy seems to talk even though the voice comes from elsewhere. Perceived taste is strongly affected by a simultaneous percept of smell — a raw potato may taste like an apple if eaten blindfolded while sniffing a real apple.

When the multisensory integration leads to wrong conclusion as when the direction of a sound is determined by vision instead of an actually correct auditory sensation we get a false percept. A true illusion arises when the multisensory integration leads to a percept originating in none of the contributing sensory modalities. An extreme example of this kind of an illusion is the McGurk effect[57]. This effect was found by Harry McGurk and John MacDonald in 1976 when they were studying visually assisted perception of speech. The face of a speaker was displayed on a monitor to facilitate lip-reading and the sound was heard from a loudspeaker. The task of the test subject was to recognize the pronounced syllable. For some reason a visual "ga" was displayed from a videotape with a simultaneous audio syllable "ba". What did the test subject hear now? Instead of the perception of "ga" or "ba" the subject heard "da". This experiment has been repeated since then many times with the same strange result. Even illusory sentences can be created around this effect. An audio sentence "bab pop me poo brive" combined with the video sentence "gag kok me koo grive" will lead to the auditory perception of the sentence "dad taught me to drive" even though the audio alone is perceived as the nonsense sentence that it really is and the video alone makes no sense either.

How is the sensory information actually combined from the contributing sensory modalities? The McGurk effect shows that the simple selection of one or the other interpretation, perhaps the stronger one, is not a satisfactory explanation as the final percept may be completely different from the available original sensory stimuli. It remains for cognitive neuroscience to find out a plausible explanation for the McGurk effect in the brain. Later on though, I will present here a simple multisensory integration mechanism that produces the McGurk effect within an artificial cognitive system.

Attention in Perception and Thinking

We seem to be free to choose what we are looking at, what we listen to and what kind of thoughts we think. We can focus our attention as we please. It is true that our attention can be caught by external events like sudden loud noises and strong visual stimuli but nevertheless it would be easy to think that basically it is the self that controls attention. Indeed, some early theories favored the concept of a central unit that would control our percepts, thoughts and actions by guiding attention. Thus "attention" would be related to "self" and in this way also to "consciousness". "Attention" might then even be a manifestation of the controlling action of "consciousness". Be that as it may, the study of higher cognition calls for the consideration of attention all the same. In what follows I will discuss the relevant aspects of attention, why it is needed, how it manifests itself, how it may operate and whether a centralized control unit is necessary or not.

The brain is a parallel processor with a huge number of neurons, synapses and connections between them. Thus its information processing capacity might also seem to be immense. Even so, the senses and memories are able to generate even larger amounts of active information. This inflow of information, if not limited in any way, would lead to the superimposition of neural signal patterns on top of each other at the same neural locations leading eventually to useless self-interfering and a contradictory neural signal cacophony. It is no use to try to listen to all radio stations at the same time; likewise the brain has to select a limited amount of information from the senses and memory for actual processing and attenuate the rest. The brain has to decide what is relevant at each moment and select the pieces of information accordingly.

This selection process is called attention here. Thus, by definition the attention process controls sensory information acquisition and the course of thought by selecting sensory information and evoked memories for further processing. It is useful to define two kinds of attention, namely *sensory attention* that is related to sensory acquisition of information and *inner attention* that is related to the selection of relevant inner representations.

Attention does not exclude non-attended information completely. This is illustrated by the so-called cocktail party effect. It is possible to attend to only one speaker against the noise of the party and yet to shift attention immediately to another stream of speech if some significant word, for instance one's own name is uttered[62,63,64]. The attention process is thus able to extract the attended stream out of the background noise consisting of the unattended sensory streams. It is clear that the attended signals will be further processed, but the system's ability to redirect attention to previously unattended streams of signals goes to show that unattended signals will also have to be processed at least to some degree.

Attentional selection may further be divided into *voluntary attention* and *involuntary attention*. Voluntary attention takes place when we consciously focus on certain sensory stimuli or a train of thought. Involuntary attention takes place when, for example, our attention is captured by a strong, novel or emotionally significant sound or visual stimulus.

What would be the neural mechanism behind attention? It can be speculated that the attended signals should be pinpointed and amplified while unattended signals should be attenuated or gated out. Indeed, attention has been compared to a spotlight, zoom lens, selector, filter, attenuator, gate, etc.[5,48,55]. Obviously the selector, filter, attenuator and gate metaphors can be easily translated into biological or engineering terms and hardware that realizes the selector function could be rather easily designed. However, the problem remains: What would control these attentional mechanisms, why should one entity be selected by these mechanisms instead of some of the others?

At each instant some information is more important or relevant than the rest. Thus importance and relevance could function as guiding criteria for attention control. For instance Donald Norman has suggested that attention is controlled by two factors, sensory activation and pertinence[6,80]. At any time certain things or signals are more pertinent than others due to context, emotional significance or general mental state and therefore on a moment to moment basis the attended items are determined by the combined sensory and pertinence scores.

From a technical point of view the simplest assumption is that attention is determined by signal intensity. Thus only the strongest signal patterns would be allowed to propagate and evoke further associations. This approach would immediately solve the basic involuntary attention control problem. The strongest stimuli, the loudest sound, the most intense light would automatically capture attention as these percepts would produce the strongest neural signals. If we assume that in many cases the strongest stimuli were also the most important at that moment then this mechanism would readily pick the most relevant sensory signals for attention. Needs can also manifest themselves as sensory signals. Hunger, thirst, etc. are detected by specific sensory processes. Thus these can also draw attention to themselves when their neural signals overpower others.

A need can thus make itself known to the cognitive system and be at the focus of attention, but the mere awareness of an active need is not enough. It is no use merely to echo in one's mind that "I am really thirsty now", one also has to find ways to satisfy the need. This in turn necessitates the focussing of attention on memories about what one has done in similar cases earlier and also on sensory percepts that would fit the evoked memories. Strong need signals may obviously evoke strong memory signals, but how could one detect the relevance of sensory signals and focus attention on them?

Sensory match/mismatch detection comes to rescue here. If a percept matches an internally evoked percept then match-condition occurs. During a match condition the sensory percept is amplified by the feedback memory signal and therefore the sensory percept becomes the focus of sensory attention. Match-condition may also be used to elevate the related thresholds, thus any unrelated lower level signals will be effectively gated out. An example of sensory match/mismatch in attention control is depicted in fig. 4.26.

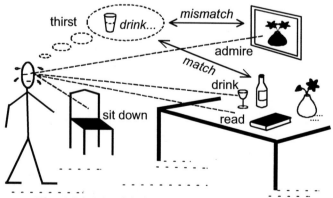

Fig. 4.26. Match/mismatch controlled attention

In fig. 4.26 a thirsty person enters a room. On which of the perceived objects should the person focus attention; a chair to sit on, a book to read, a painting to admire? All these generate a mismatch against the inner imagery that is evoked by the active need, thirst. Thus these percepts are not relevant to the need and therefore attention must not be sustained on them. Eventually visual attention may be focussed to the bottle on the table. This percept will now match the inner imagery and the focus of sensory attention shall be sustained so that the bottle evokes further memory images, mainly those that are related to the act of drinking.

This example illustrates the role of match/mismatch conditions on attention control. In general it can be seen that the cognitive system should strive towards match-condition. Thus the desired effects of the match/mismatch/novelty conditions on attention can be summarized as follows:

Match	$\Rightarrow$	sustain attention
Mismatch	$\Rightarrow$	refocus attention
Novelty	$\Rightarrow$	focus attention

If we assume that the strongest signals attain attention automatically then the desired effects of match/mismatch/novelty on attention may be achieved by low-level signal strength and threshold control. All this can take place without any higher level supervisor or controlling agent.

However, attention control by match/mismatch/novelty conditions does not cover all situations. In more general cases the significance of available percepts must be evaluated and attention must be focussed to the most important of these. An example of attention control by significance is depicted in fig. 4.27.

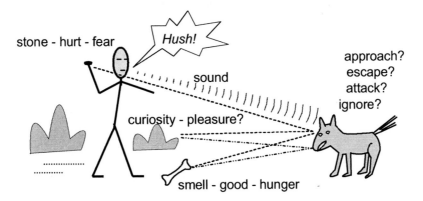

Fig. 4.27. Attention control by significance

In fig. 4.27 a hungry dog finds a bone, but there is a twist. There is also a threatening man there with a stone. Other percepts are also available for the dog; there are interesting bushes that evoke curiosity. This situation presents a serious attentional problem for the dog as improperly directed attention may lead to painful consequences. Obviously the significance of each percept here can only be evaluated by experience. Thus the dog must have memories of men and stones, the pleasures of strolling in bushes, the taste of juicy bones. The focussing of attention on one of these would evoke action that would exclude other actions for that moment. In the previous example attention was guided by signal strength. The same mechanism can be used here if the significance of these percepts is carried by the strength of the corresponding signals. Thus the most significant percepts would also be represented by the strongest signal patterns and these in turn would automatically become the focus of attention. In this example the percepts of the shouting man might have the greatest significance. Therefore the dog's attention would be directed to these and possibly the act of escaping might be evoked. On the other hand strong hunger might override the significance of the threatening man and the dog might try a dash for the bone. Here again attention is controlled rather automatically by the relative significance of instantaneous internal needs and external percepts.

These examples try to illustrate how attention may be guided by situations, needs and learned subjective significance. The focussing of attention on any specific object may also lead to corresponding action. There is no need for any higher level supervisor, no special "attention box" is needed as the attention mechanism is distributed within the neural system. I would even say that there is no specific attention mechanism, the function that we call attention is simply a biological neural system's basic way of favoring the strongest signals, a process that is present already in the simplest central nervous systems. Nevertheless, this is not to say that in complex neural systems like the human brain certain brain areas could not contribute to attention more than some others.

Perception of Self as a Vantage Point

Look around and you will see that things are located out there and you can even estimate their distance from you. Did you hear something? Now again the source of that sound seems to be out there and you can even tell more or less accurately the direction of the sound. You are a moving vantage point in the middle of everything else out there. This vantage point of view makes us a localized observer and therefore is a key element of our sense of self. This is a feat that we seem to accomplish effortlessly, yet there are some crucial questions there. The first question is, how is it possible to perceive the origination point of a sensation as being somewhere other than the point of the sensor, the tip of the nerve, the real point of origination of the nerve signal. Secondly, how can a mental map of the surroundings be created. Thirdly, how can the distinction between the percepts originating from the external world and those generated by our own mental activity be made.

The first question relates to the apparent point of origination of a sensation. The point is, what we actually see is only the projected image on the retina, what we actually hear is the vibration of the eardrums, so why don't we just perceive these as such, percepts originating at the senses or originating at the related sensory nerve endings? How can we possibly perceive the situation being anything else? But as we do, does it follow that the trick that our brain is doing must be extremely complex?

Let's consider a simple experiment to illustrate the point further. Let's scan a rough surface with a rigid stick. What do we feel? It so happens that we do not perceive the vibrations of the stick against our fingers, instead we perceive the groove patterns of the surface. There are no nerves on the tip of the stick, so what is happening?

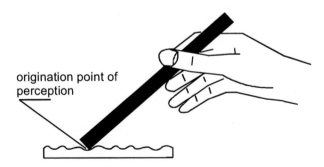

origination point of perception

Fig. 4.28. A stick on a rough surface

Since there are no nerve fibers going through the stick and into our brain common sense would say that we could only sense the vibrations of the stick against our fingers. In fact this is exactly what is happening, so why do we perceive things to be different, why do we perceive the rough surface out there?

Let's consider a doorbell as an analogy. A doorbell system consists of the actual bell, a battery, a push button at the door and a length of wire that connects these components.

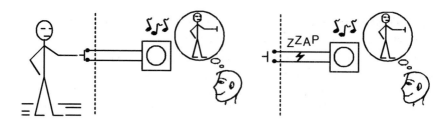

Fig. 4.29. The doorbell effect. Is there somebody at the door or is there an electrical short circuit? Both cases evoke the same mental image.

If the doorbell rings we know by experience that someone is at the door pushing the doorbell button, we may even have a mental image of this. Now suppose that an intermittent short circuit somewhere along the wire makes the bell ring. What would be our first impression?

The ringing of the bell does not convey any information about the point of origination, therefore our impression will again be that someone is at the door. The meaning of the ringing of the bell is an association, learned by experience, not something that is inherently wired in the system. Naturally rigid wiring is necessary. If we had several doors, bells and buttons and the connections from the buttons to the bells were randomly changing all the time, we could not make any sense of it and whenever a bell rang we would not have any idea to which door to go.

In the same way nerve signals do not carry any position information about their actual end-location even though they are rigidly wired. From the brain's point of view the nerve signals just appear, there is no built-in point of origination. Therefore the perceived point of origination can be whatever we associate it with. The rigid stick can convey the sensation of the rough surface. In early childhood we learn that a certain visual sensations correspond to tactile sensations of objects out there and to the motor commands that enable us to reach them. In this way we associate sensations from the sensory modalities and motor commands and, thanks to the rigid wiring, will eventually come to develop a coherent understanding that the sensory signals depict entities out there. We have to reach out for them, not to our eyes or ears. The positions of the actual nerve endings at the retina, ears, etc. will not matter, they will be transparent to the system. The origination points can appear to be out there, this is not achieved by design but by the omission of nerve end position information.

Thus the point of origination can be associated rather freely with a sensation. Due to this freedom it is possible to associate a direction and distance to visually and aurally perceived objects by the mechanisms already explained. These mechanisms evoke signals that correspond to the direction and distance of an object. The perception process must now combine and bind together the actual visual and auditory percepts of the properties of the object as well as the related direction and distance percepts.

The vantage point of view arises from the dynamics of the object position and distance perception. The perceived object positions are relative to the observer's own position and reference direction, and will change when the observer moves his head or changes position or when the objects themselves are moving. The position information of the external objects allows the short-term memorization of a kind of a mental map of the immediate surroundings. We know what is where relative to us, we have an idea what is behind us even though we cannot see there without turning our head. We know what it would take to reach out for these objects.

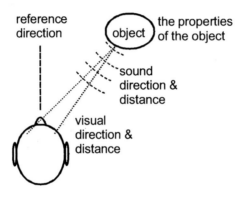

Fig. 4.30. Visual and auditory perception of an object out there

However, the direction and distance information are not responsible alone for the perception that the sensation originates from outside. This fact can be demonstrated by the "headphone illusion", fig. 4.31. When monophonic music is heard via headphones the apparent point of origination of the sound seems to be inside the head, midway between the ears. Yet we know that the actual source of the sensation is external, we know that the sound is not generated within our head.

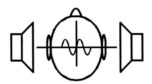

Fig. 4.31. The headphone illusion, the sound is inside the head

Obviously there are at least two mechanisms that allow us to see that the sound in our head has external causes. First, the corresponding activity at the auditory sensors points outward to external stimuli. Second, we can recognize that the sound is not produced by us, it is not our auditory imagination. Of course it is possible silently to hum the heard tune. In this case however, the hummed tune is distinct from the actually heard sound and can be perceived as such, as our own product.

Thus, the vantage point of view arises from the ability to attribute to sensations a point of origination that is different from the actual sensory nerve end position, the ability to detect visual and auditory directions and distances, the ability to create mental maps of surroundings, and the ability to distinguish between externally and internally generated sensations.

CHAPTER 5

LEARNING, REASONING AND INTELLIGENCE

Learning, Memorization and Memory Making

Basically, learning is the process of acquiring knowledge and skills. Some psychology books give it a fancier definition like a process that changes the subject's behavior. I will not use that definition here because not all learned knowledge changes behavior and because this kind of a definition would obviously also include forgetting.

There are a number of learning mechanisms. The most primitive forms of learning include habituation and sensitization[103]. Habituation occurs when a system's response decreases when the stimulus is repeated. For instance we learn not to pay attention to the noise of a fan or other repeating background sounds. However not all repeating stimuli lead to habituation. The background noise may be so intense that day after day we get more and more irritated by it. This is sensitization, a system's increased response to a repeated stimulus.

Associative learning is a more advanced mechanism that allows complex links between stimuli, representations and responses[65,104]. Associative learning mechanisms include correlative learning, learning by trial and error, learning by imitation and learning by description.

Learning by heart is one form of associative learning. This involves the memorization of information as such like "Mary had a little lamb....". It is also possible to memorize contents of imagery. With the word "memorization" I will also refer here to the act of imprinting something into memory. By "memory making" I refer to the making of episodic memories of our own personal history. There is a clear difference between memorization and memory making. Memories of significant events are made automatically on the run, while memorization usually involves repetition, rehearsal and mental effort.

Sometimes a distinction is made between semantic and skill learning. Semantic learning involves the learning of a fact, like "a rose is a flower". Skill or procedural learning involves the learning of skill routines, mental or motor sequences.

Learning usually involves some kind of adaptability and generalization, the ability to use the learned information inductively in different situations. We don't have to learn driving anew each time we have a new car or enter unknown roads. Learned facts and skill routines are not usually accompanied by memories of exact learning episodes, we do not necessarily remember how and when we learned that "a rose is a flower".

Associative Learning

Classical or Pavlovian conditioning is a simple example of associative learning. Russian physiologist I. P. Pavlov discovered the so-called conditioned reflex effect in 1901 while doing experiments with dogs. Pavlov noticed that if a bell is rung each time when a dog is given food then the dog becomes conditioned to the sound of the bell; if the bell is rung the poor dog will salivate even though no food is present. This conditioning is not permanent. If the ringing of the bell is not accompanied by food repeatedly, the salivation reflex will fade away. Other stimuli, like light, smell, touch, etc. may be used instead of the sound of the bell.

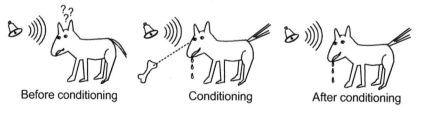

Before conditioning   Conditioning   After conditioning

Fig. 5.1. Pavlovian conditioning

In Pavlovian conditioning we have a so-called unconditioned stimulus, in this example food, that always leads automatically to an unconditioned response (salivation). During conditioning a second stimulus (the bell), which does not by itself lead to the unconditioned response, is associated by repetitive temporal contiguity to the unconditioned stimulus–response chain. After a while the new stimulus will be able to elicit the unconditioned response by itself without the presence of the original unconditioned stimulus. The new stimulus is now called the conditioned stimulus.

The chain of associations can be extended by conditioning another stimulus to the conditioned stimulus so that the new stimulus by itself will again be able to elicit the original unconditioned response. This is called second order conditioning[65].

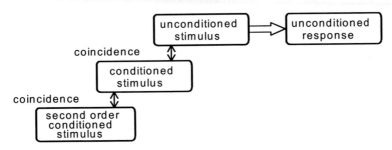

Fig. 5.2. Classical first order and second order conditioning

Associative learning that is based on simple conditioning by temporal coincidence is not the whole story. In many cases the action and effect or other entities to be associated do not occur at the same time. For instance a rat may eat poisoned food, become ill much later and yet be able to learn to avoid that special food in the future. Stomach pain and the taste of the poisonous food do not occur simultaneously, so how could an association be made with direct conditioning? Examples like these have been used to argue that associative learning is not the answer except in cases of simple conditioning[86]. However, associative learning may still be possible if a system operates with inner representations that can be evoked later whenever needed. In that case associative connections would be established between the inner representations instead of actual sensory signals. Then for instance a stomach pain might evoke the memory representation of the last consumed food and the pain could be then associated to that specific food via this recalled memory representation.

However, associative learning has other problems. A cognitive system may have numerous representations active at the same time, some originating via sensors from the environment, others evoked from memory or initiated by internal sensors. Which representations, if any, should be associated together? Without any kinds of constraints the establishment of useful associations would seem to be unlikely.

Attention can be used to limit the number of signals so that only relevant associations arise. For example we may teach the name of an object by focussing the system's visual attention on this object alone by pointing it out and at the same time giving the intended name (a table, a pen, etc.). This can be done if the cognitive system's visual system has a definite small focus area, something like the fovea of the human eye, so that the direction of the gaze can be used to pinpoint objects. In this way only the attended items, in this case auditory and visual, will be associated together.

Correlative Learning

Basic associative learning can connect two entities together, for instance an object and its name. Unwanted associations can be avoided by focused attention as described above.

However, there are properties that cannot be pointed out separately and therefore an exclusive one-to-one coincidence cannot be presented. Sizes, shapes etc. do not appear alone, they are properties of objects that also have other properties. For instance, suppose that we want to associate the name for a shade of an object (e.g. "grey") to a cognitive system. We could show a grey object and give the word "grey". But how would the system know that for which property of the object the name was intended? How could the system know that we do not intend to say that the object's name was "grey"? It really does not, so how do we do it?

Correlative learning is one answer to this problem. We can use several different objects that have the property to be named in common. Then we proceed as usual and give the name of the intended property in context of each of these objects. Now the name will be associated to the desired property every time an association is made because every object has this common property. However, the name will also be associated to the other properties of these objects, but in the end not so strongly as to the common property.

Fig. 5.3 gives an example how the word "grey" can be associated to the corresponding property by using three grey but otherwise different objects.

Association pair	●	△	 □	 ○
"grey"  & △	+1	+1	+0	+0
"grey" & ■	+1	+0	+1	+0
"grey" & ●	+1	+0	+0	+1
Association strength	3/3	1/3	1/3	1/3

Fig. 5.3. Correlative learning; giving a name for a common property

In fig. 5.3 the name "grey" is associated to the respective property with three example objects, a grey triangle, a grey square and a grey circle. At the end of the training session the name "grey" is associated to the property <grey> with the cumulated strength of 3/3 while the undesired association strengths to the other properties remain at 1/3. This is already a good state of affairs. This situation could be further improved by using threshold controlled association strength decay so that only a strong associative links that exceed a given threshold would remain while weaker links would decay away. In this way after a while only the intended association would remain.

Thus correlative learning solves the problem of general property naming by cumulative association and forgetting. We need to forget in order to be able to generalize.

Learning by Trial and Error

What kind of action should be executed so that a desired result can be achieved? If we do not know the answer, we can always try something and see if that would result in the desired outcome. Then, for later occasions we could memorize the successful approach.

This leads to learning by trial and error[47,88]. Such processes are sometimes called instrumental or operand conditioning[66]. Obviously desired outcomes lead to pleasure and are thus rewarded while undesired outcomes lead to displeasure and punishment. In this way learning by trial and error will be directed by positive and negative reinforcement and successful sequences of actions may be eventually learned.

The trial and error method is not generally very efficient, because the number of wrong ways to do something may be very large while only one or few ways will lead to the desired outcome. Thus an excessively large number of trials might have to be executed. However, the number of necessary trials can be drastically lowered if the results are cognitively evaluated. One can learn also from failures and shortcomings; the evaluation of the reasons for a failure may give clues about the successful approach. In this way the most promising line of trials can be identified and the number of required trials reduced.

Learning by Imitation

Why go to the trouble of stumbling through trials and errors when others have already mastered the art? Let those who know show you how it is done or simply copy them. Social or observational learning is a more efficient learning method as it allows the copying of others' expertise and therefore largely bypasses the need for one's own trials and errors[67]. In this form of learning the subject observes the actions of others and imitates them as such or in a modified way. Nowadays learning by imitation may obviously utilize movies, television and other audiovisual media.

Not all imitation leads to meaningful learning though. Little children and some animals imitate the actions of others without understanding the purpose or the consequences of the action, sometimes with unfortunate outcomes. Even though the act itself may be learned as a mechanical sequence, it does not constitute meaningful learning, as it has not been associated with the achievement of any goal. However, this playful imitation may be a necessary precursor for the future ability of meaningful learning.

Real meaningful learning by imitation involves the realization of the purpose and consequences of the act and the understanding of the relationships and relevance of the different phases of the act. The subject must realize why something is done, the way it is done and how this is related to the goal; "my need is this and — aha! — I can achieve that in this way".

Two different methods of observational learning may be distinguished, namely (1) direct imitation and (2) observation with cognitive judgement. By direct imitation one reproduces the observed action as such or with very little change. By observation with judgement one may learn not to do what others have done.

Observation with judgement involves the evaluation of the observed action against one's own motives and the mental evocation of expected rewards or punishments. In this way the observed action may be assimilated in similar style to direct imitation or it may be avoided in the future.

Direct imitation, for instance the learning of a set of motor actions, involves the observation and association of correspondence between one's own motor actions and those observed in others. Thereafter learning involves the serial association of one's motor actions so that the observed motor sequence can be performed.

There is a neurological basis for direct imitation. Magneto-encephalography measurements have shown that when a subject sees an action being executed by others the same neuron groups that would be responsible for the execution of the seen action will be activated[43,79]. Even the activation order of the neuron groups is the same during observation and actual execution. In this way there will be neural readiness to reproduce and imitate the seen act. These neurons especially in the so-called Broca's area at the left hemisphere of the brain are called mirror neurons, as their operation would seem to mirror the observed actions of others. It has been speculated that speech learning could be closely related to this mirroring action because Broca's area covers neurons responsible for speech production and also mouth and hand movements. Mirror neuron action was initially found in monkeys by Rizzolatti[84].

Learning by Verbal Description

Another form of social learning is the indirect learning by spoken or written description. In this case there are a teacher and a learner who share a common communication system, a language. The task of the teacher is to produce proper mental imagery and ideas in the mind of the learner and have them permanently associated in the memory. Correct imagery can only be evoked and new material successfully taught if the learner has already the basic concepts that are needed. The learner must understand what is being taught, he must be able to evoke the relevant connections, concepts and ideas. Therefore teaching must be hierarchic, new concepts cannot be learned if there is nothing that they can be associated with.

Hands-on trial and error learning and imitation learning rely on a real world situation where the environment can be used as a memory aid. Objects and their relationships are available continuously and attention can be brought to bear on their details when needed. On the other hand verbal description relies heavily on limited capacity short-term memory. This can be compensated by the use of external aids like drawings, pictures, blackboard, etc.

Written language enables learning from books. However, a real teacher is better because of the feedback possibility. The learner may ask questions whenever understanding fails, the teacher may monitor the emotional expressions of the learner to see for instance that attention is paid properly. Face-to-face contact also enables the transmission of emotional cues and the mirroring effect so that the learner may better follow and anticipate teaching.

Memory and Memories

There can be no learning without some kind of memory. Memory is also necessary for deduction, reasoning, imagination, planning and the execution of these plans. Deduction, reasoning and mental arithmetic necessitate the short-term storage of intermediate results. Skills involve the execution of definite mental or motor steps in a certain order and therefore need memory. Different cognitive processes need different kinds of memory. It is useful not to forget acquired skills. On the other hand it may be useful to forget information that is not needed, especially if the forgetting frees memory resources and streamlines the cognitive process. In certain cases forgetting is absolutely necessary. For instance the current referred meaning for words like "it", "this", "you", etc., must be released soon otherwise they would eventually actively refer to ever increasing number of objects.

The memory in the brain is not like the computer memory that stores data at addressed locations. There are no specific memory areas in the brain and addresses are not used to locate information. Instead the memory function seems to be an integral part of the cognitive process that is distributed all over the brain. Therefore, in the following when the memory in the brain is discussed the distributed memory process is intended, not a localized store.

In cognitive psychology the memory is often divided into the following components: short-term or working memory, episodic long-term memory, semantic long-term memory, procedural or skill memory[7,30]. In addition there is evidence that there exist sensory memories that store the sensations in rather unprocessed forms for a little while like the echoic memory for sounds[68].

The short-term or working memory is responsible for the immediate availability of the representations that are needed for the on-going mental task. This memory has very limited capacity, usually cited as 7 ± 2 items[8]. Short-term memory is necessarily volatile, due to the limited capacity and the fact that temporary associations must be disassembled. Possible neural memory mechanisms for short-term memory could include sustained neural activity in the form of one-shot behavior of neurons where short synaptic triggering leads to prolonged firing[37]. Another mechanism could be based on reentrant neuron loops where feedback would sustain firing patterns. One possible major type of this kind of a loop may be the thalamo-cortico-thalamic reverberatory circuit[24].

The episodic long-term memory is responsible of the episodic memories, strings of events, what happened, including system's own history. Episodic memories may appear as short movies in our heads, but in reality they are hardly stored in that way. More probably they are reconstructions that are assembled during recall. Episodic memories are made automatically during our life, but the episodes to be memorized permanently must be significant in some respect. It is also possible to make episodic memories through rehearsal or stimulus repetition. We can learn poems by heart, we will learn musical pieces by repeated listening. The serially associative nature of these memories is easily detected. For instance when we repeatedly listen to a musical record, say a pop album, we will eventually notice that the first notes of the next piece will come to mind during the last moments of the currently playing piece. This I call the LP-effect after the long-playing records. Serial association operates also in learning by heart. When I was in the elementary school it sometimes happened that one was not able to remember all verses of a poem. The teacher's advice was: Start again from the beginning and get speed, the rest will follow. A new start is good because it re-establishes associative links and speed is good because attention is less easily diverted. Attention diversion may also be avoided by looking upwards so that visual stimuli are minimized.

The semantic long-term memories contain semantic knowledge but without any information about the actual episodes when that information was acquired. Also these memories seem to operate associatively; recalled entities seem to be evoked by those in the short-term memory.

Long-term memories seem to be stored in the neural circuits as learned transmission strengths of synapses. This mechanism is also called the associative long-term potentiation (LTP)[35,76,99].

How are memories recalled and perceived? The mere associative activation of some odd group of neurons somewhere in the brain does not necessarily amount to the perception of memories, therefore some mechanism that can bring the evoked representation into attention must be proposed. I am proposing here that recalled memories are perceived by the same process that is used for sensory perception. This can be achieved with a feedback that projects the evoked neural activity patterns back to the sensory perception area. Reference is made here to the chapter on perception.

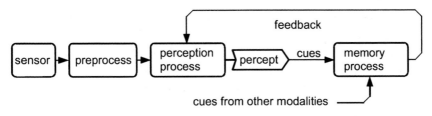

Fig. 5.4. The perception of memories

A memory pattern is evoked as a response to a cue. Many kinds of cues may be used, our own thoughts, something we see, hear, even smell, whatever. The evoked memory pattern is then routed back to the perception process where it will be translated into a percept. The feedback loop may sustain the memory percept for a while and thus act as a short-term memory. This architecture will also allow the evoked memory to act as a cue for further memories.

What is the difference between memories and imaginations? Imaginations are not necessarily visions of things never seen before. I can imagine drinking espresso, an act that I have actually done many times before. However this imagination is not necessarily a memory of such an occasion. I can imagine picking up a book on the table. The components of imaginations may originate from actual sensory percepts and from memory. We can imagine a seen object to perform some act, we can imagine what to do with the object. In fact we are continuously connecting remembered actions to seen objects mentally and eventually some of these actions get realized. Memories and imaginations arise from the same neural areas and are perceived in a similar way, therefore confusion would seem to be possible.

What are memories for from a practical point of view? They are there so that we can utilize experience. The utilization of experience takes place via imagined action based on memories. Here memories with associated good/bad values are only raw material and therefore their faithfulness is of secondary importance. Thus it seems to me that the ability to distinguish imaginations from true memories is just a luxury, a function not really needed. However, we seem to be able to make that distinction, at least to some degree, so what could be the contributing factors?

True memories are usually more vivid, they contain more details and are associatively connected to other memories. However, it is possible to rehearse imaginations so that they attain vivid details and if these imaginations were now supposed to depict an episode that took place long time ago then it may be very difficult to distinguish them from real memories. Various experiments have shown that imagined childhood events can turn into false memories[54]. The more a suggested childhood event is imagined, the more vivid imagery is created and eventually these can be taken as true memories. On the other hand real memories must be consistent, they must not contradict each other. Imaginations that contradict memories generate mismatch conditions and are therefore recognized as what they are. Real memories must not contradict external evidence either.

Perception of Time

We perceive time because we perceive events and accumulate memories of these, our personal history. Time is going forward because today we have more memories than yesterday and we know that today's memories are more recent than yesterday's.

The perception of time involves three different aspects: (1) the episodic sense of time, (2) the ability to estimate long and short intervals and (3) the perceived speed of time. Our perception of time relies very much on memories and retrospection and also on our expectations of the future. These are represented in the brain basically in the same way as the present moment, therefore the processes for the perception of time must also include mechanisms that allow the distinction between the representations of past, present and future.

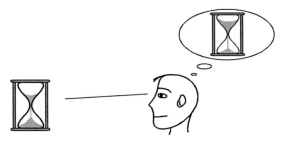

Fig. 5.5. We perceive passing time because we can make memories and note the change

The episodic sense of time arises from our organized memories. We have memories of the past, we know what we did just a while ago, an hour ago, yesterday, last year and the time before that. We have a sense of recentness, we know if these things happened recently or long time ago and in which temporal order. We have also an expectation about what is going to happen next, what we are going to do in the future.

How old is a memory, did the remembered episode take place just now or a long time ago? Recent episodes are represented by vivid memories that are rich in detail. Older episodes are represented by fainter memories with fewer details and eventually the memory trace is so thin and sketchy so that sometimes one wonders if those things happened in that way at all.

The temporal order of recent events is most probably deduced directly from the episodic nature of the memories. This is also true for episodes that took place a long time ago but what was the order of these episodes? It seems that no temporally ordered day-to-day and episode-to-episode linking is preserved, therefore the order of past episodes is not so clear. We may have difficulty recalling in what year exactly a certain event took place. The vividness of the memories may give some cues of the temporal order, but then some old but emotionally very significant memory may retain its vividness and the perceived temporal order can be distorted. In these cases the temporal order of old memories may only be resolved by causal before–after deduction.

How is the temporal duration of a period estimated? We are not able to perceive the passage of time directly, mainly due to the fact that there is nothing to be perceived because time as such does not exist. Therefore we must link our concept of time to real events and their relative order. The measuring of time involves the comparison of two or more sequences of events. How long does it take to breathe in and out? Take a clock and find out. In this case the clock provides us with the reference sequence of events, the numbered ticks, while the breathing is obviously the other sequence of events. Thus in general the perception of time must be based on the perception of various actual processes set against others. We are not born with wristwatches but there may be some biological processes that the brain is using instead.

The estimation of short intervals might be based on the comparison to the speed of neural activity, speed of verbal thinking, rate of breathing, heart beat rate, etc. However it may be possible that the brain contains specific timing circuitry for this purpose.

The perception of the duration of longer intervals can be based on internal processes that involve an increasing or decreasing variable like hunger, thirst, fatigue and boredom.

The perception of the duration of even longer periods may be facilitated by external cues like day rhythm, seasons, and artificial means like clocks and calendars.

The biological daily rhythm is known as the circadian rhythm. Human body temperature, blood pressure, hormonal balance, sleep and wakefulness follow a rhythm that is usually synchronized to the 24-hour day. However, experiments where subjects were deprived all external cues of time seem to show that the innate circadian rhythm is a little bit longer, about 25 hours. (Also our everyday observations seem to confirm this, many of us would like to sleep another hour in the morning!) The innate rhythm is obviously a product of evolution, how come then that the million-year evolution has not been able to get the rhythm right?

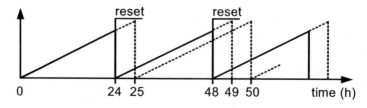

Fig. 5.6. The synchronization of circadian rhythm by daily resets

Now every electronics engineer realizes immediately that the innate rhythm is right and must have a period longer than the 24-hour day. Obviously the innate rhythm is synchronized to the 24-hour rhythm by a reset every morning. If the innate period was shorter, it would reset itself before its due time and synchronism could not be maintained.

Why does time sometimes seem to go by so slowly and then other times so fast? Does time have a speed? The concept "speed of time" is a bad misnomer, because in physics "speed" is defined as something happening in a unit of time. Time divided by itself does not yield the "speed of time". However, we can always compare two clocks and determine that one of them goes "faster" or "slower". In the same sense we can compare our subjective sense of time, our "mental clock" to external conventional clocks and determine the speed of our "psychological time". Thus in the following I will use the concept "speed of time" to depict the sense of the subjective duration of given time intervals.

When we are sitting in the dentist's chair or waiting for a boring lecture to end time seems to go by so slowly. Then again, when something pleasant happens, time may seem to go by much too fast. When we have nothing to do for an extended period like on intercontinental flights the time barely goes by. This is generally the case when you do nothing but only wait for something to happen. When you are young, the days seem long and the years even longer, when you are older the days get shorter and so do the years, but when you reach old age the days will be again long but the years are even shorter. Is there something here or are these only anecdotes? How could the perceived speed of time be subjective?

When young kids have nothing to do and just wait for something to happen, they keep on asking: Is it time? No, not yet. This is true for adults also, if you wait for something and keep watching the time, time freezes. The perceived time does not go forward because you are stuck to the one percept: Is it time? — Not yet. Time seems to go by because the comparison between present and previous percept shows that something has happened, but in this case the two percepts are the same, there is no change and no illusion of passing time arises. So, when nothing happens, time seems to go by slowly.

How about the speed of time in retrospect then? Why would the past days, weeks, months and years seem short or long? This can be explained by the way that the brain records memories. The brain is not like a tape recorder and our memories are not like a tape. When a tape recorder records for an hour, the length of tape for an hour is consumed be it sound or silence. Afterwards it is easy to find out how long a given episode lasted, just measure the length of the consumed tape. The brain records only if something worth recording exists. Therefore the size of the record, the memory space or the number of associative links for a given time interval will not be constant and in retrospection will not give a consistent estimate of the length of the period. In childhood all things are novel and worth memorizing, lots of memory space is used and a large number of associative links are formed for the early years. Likewise a holiday abroad may seem long as novel things happen. Afterwards the large number of associations gives us an illusion of long days and years. Later on the actual memories may fade away but the memory of the time illusion remains. In old age everything is already familiar, there is no need to note these. Not much happens either, time goes by slowly in waiting for the few highlights of the day; a meal, perhaps a favorite TV-show. During these uneventful days only few associations are formed and minimal memory space is used. Therefore, even though the days go by so slowly, the years seem short in retrospect.

Our memories, present moment and expectations of the future are all represented in the same way, by similar neural activity patterns in the brain. These representations must be distinguished from each other, though, otherwise we might respond to memories and expectations as if they were happening now, at this moment. How do we achieve this distinction?

First of all, representations that depict present moment originate from senses, outside the brain. Internally evoked representations do not have associated sensory activity and are not therefore referred to outside sources. Also the mental representations that are supposed to represent the present moment must be compatible to the actual perceived situation. If I now were to evoke memories of being in Africa then just a quick look around would convince me that these mental representations would not represent the present situation. Thus cues from the environment help to resolve the distinction between memories and representations of the present.

What would happen if a representation of the present moment temporarily lost the connection to its sensory origin? Obviously this representation could then no longer be distinguished from those originating from memory. When a fraction of a second later the same representation is again received from the sensors it would match the "memory representation" and an instant sensation of familiarity would be evoked. We would get the *déjà vu* experience, I have seen this before.

We perceive time because we accumulate memories, our personal history in the ways outlined above. A machine can be given a similar sense of time, not by providing a clock but having it also accumulate associatively connected episodic memories like we do, by letting it create its own personal history.

Deduction and Reasoning

The perception process creates percepts of what is directly observed. Visual objects, sound, whatever is sensed, are transformed into respective inner representations. However, in many cases the direct observations are not enough, the required information can only be achieved indirectly by deduction and reasoning. Artificial Intelligence uses formal logic and rule based reasoning. Humans can also reason by learned or discovered rules, but all human reasoning is not based on explicit rules. How does one deduce and reason without rules and logic?

Black clouds gathering in the sky — what does it mean? The direct observation leads to the inner representation of the clouds, no more.

Fig. 5.7. Simple deduction by evoked experience

By experience it can be deduced that it may soon rain. Simple deduction involves the prediction of consequence to a condition and this can be done directly by associative evocation of previous experience.

How do we get rules and logic? Experience leads to the observation that some entities are causally connected, one entity being the cause, a necessary prerequisite for another. However, causality is not that simple and the mere association of the cause and the consequence is not enough here. A single association can only connect two entities together, for instance rain and clouds. From this association it can be inferred correctly that if there is rain there must be clouds. But also a wrong inference can be made: If there are clouds there is rain. Additional experience and associations are needed to establish the fact that there may be clouds without rain and clouds may only indicate the possibility of rain. The ability to entertain several associations at the same time is needed for correct deduction by causal connection.

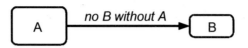

Fig, 5.8. Causally connected entities

A causal connection can be formulated as "no consequence without cause" or "no B without A". This can be written also in the form "If B then A". This latter form is conditional and thus this mode of reasoning is also called conditional reasoning.

The following conclusions can be derived from the causal connection:

There is no B without A	There is no B without A
There is no A	There is B
Therefore there is no B	Therefore there is (some) A

If it rains there are clouds	No smoke without fire
There are no clouds	There is smoke
Therefore there is no rain	Therefore there is fire

What else could be deduced? Can we deduce that if there is A then there is B? No, there may be clouds in the sky without rain, there may be fire without smoke. Can we deduce that if there is no B then there is no A? Again no, as before this situation says nothing about A. We can see that there are four possible inferences of which only two are valid.

no cause present	– no consequence present	(valid)
cause present	– consequence present?	
no consequence present	– no cause present?	
consequence present	– cause present	(valid)

Another everyday form of reasoning, deduction by exclusion is used frequently. If we eat the cake today, we know that it will not be there tomorrow. If we go to the concert then we cannot go to the movies. If we choose one possibility the other ones will be excluded. This reasoning style has the following form:

A is either B or C	A is either B or C
A is B	A is not B
Therefore A is not C	Therefore A is C

Two examples:

John is either at the library or at movies	Bill is either studying or sleeping
John is at the library	Bill is not studying
Therefore John is not at movies	Therefore Bill is sleeping

Deduction by exclusion provides an alibi. A person or a thing cannot be at two or more locations at the same time. Also some actions are exclusive; you cannot both spend your money and have it for future use. Multiple premises and choices result in more complicated forms of deduction by exclusion like in the following example:

| A is either C or D | Tim is either six or nine years old |
| B is either C or D | Tom is either six or nine years old |
A is not B	Tim and Tom are not of same age
A is not C	Tim is not six
Therefore B is C	Therefore Tom is six

Causality and exclusion do not cover all reasoning. It is also possible to deduce by the relationships between entities. Deduction by relative magnitudes goes like this:

Ice cream is better than milk	Gold is more expensive than silver
Candy is better than ice cream	Silver is more expensive than bronze
Therefore candy is better than milk	Therefore bronze is cheaper than gold

Three entities are compared to each other here. The inferences from this kind of comparison are as follows:

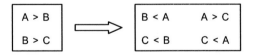

Fig. 5.9. The inferences from the comparison between three entities

However, watch your step here, what can you make of this?

Ice cream is better than milk	Gold is more expensive than bronze
Candy is better than milk	Bronze is less expensive than Silver
Therefore ice cream is better than candy?	Therefore silver is cheaper than gold?

Here the two premises do not carry the information that would be needed for the proposed conclusion. In the first case the premises A > C and B > C do not tell anything about the relationship between A (ice cream) and B (candy). In the second case the premises are A > B and B < C. However B < C is equivalent to C > B and again we can see that no information about the relationship between A (gold) and C (silver) exists. Therefore the conclusions are logically invalid even though they might sometimes be true in practice.

Deduction can also be based on comparison between categories or classes. Syllogisms are three-statement logical forms that do this. The first and second statements of a syllogism set the premises and the third statement is the conclusion. There are several ways to illustrate syllogisms graphically. Some examples of syllogisms are given here.

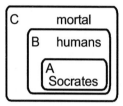

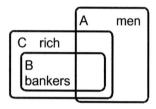

All A are B
All B are C
Therefore all A are C

Socrates is human
All humans are mortal
Therefore Socrates is mortal

Some A are B
All B are C
Therefore some A are C

Some men are bankers
All bankers are rich
Therefore some men are rich

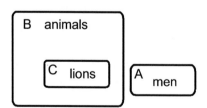

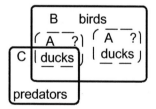

No A are B
All C are B
Therefore no A are C

No man is an animal
All lions are animals
Therefore no man is a lion

All A are B
Some B are C
Therefore some/all A are C?

All ducks are birds
Some birds are predators
Therefore some/all ducks are predators?

The last example shows an invalid inference. The diagram shows that the position of A, the class of ducks, cannot be determined by the given information, therefore the suggested inference cannot be made.

Modes of reasoning like the methods presented above can be formalized exactly because the processes do not introduce any new information. This is the essence of logic; same is same. Logical operations do not change the original information. Only the way that the information is represented is transformed so that the desired facet of the information could be perceived; so that it could fit into the limited working memory space in our brains and become the object of attention. This kind of reasoning would be quite unnecessary for a supermind with unlimited working memory. For instance consider the question "are some men lions?" We could start with the logical reasoning chain: No A are B, All C are B, therefore no A are C, then put men, animals and lions there and eventually come to the conclusion that no man is a lion. Alternatively we might pretend to be superminds and look at the respective picture above that contains the very same information in graphical form. This representation allows the instantaneous access to all relationships and we can see immediately that no man is a lion. So, if we could visualize problems better, see all information in a parallel way, we would not need to transform representations and utilize rules of logic.

A large part of our reasoning is made with incomplete information. In those cases extra information is introduced by the reasoning process, this additional information being the result of our subjective decisions, our own wisdom (if any). This kind of deduction may be based, for instance, on preceding examples and analogy. What has been experienced in similar cases before may indicate what should be expected now. Proper analogy-based reasoning involves the identification of a known case whose relationships can be carried over to the present case. If we have learned that two apples plus two apples make four apples all together then we can carry this relationship over to two and two bananas.

Generally reasoning with incomplete information involves the evocation of several mental scenarios, detection of contradictions, evaluation of their emotional significance and the rejection of alternatives with foreseen unsatisfactory outcomes. Shortcuts may be used like the selection of the most familiar. Absolutely correct outcomes cannot be guaranteed and different people may draw different conclusions in the same situation.

Intelligence

What is intelligence? What is typical to intelligent behavior? On what mechanisms is intelligence based? These are crucial questions to anybody wishing to create artificial intelligent machines. Unfortunately present day psychology and cognitive neuroscience do not give complete answers to these questions.

Let's consider a practical situation where intelligence is required. Suppose that you have acquired a novel gadget, one of those absolutely necessary today and you wish to learn to operate it. No intelligence is needed if the gadget can be operated by exact and clear rules, just follow the instructions given by the owner's manual and everything will turn out right. However, if the instructions were so unclear that you had to guess and figure out what they mean then you would need intelligence to use that gadget. Here you have the problem; what do the instructions mean? Eventually you may figure out in one way or another how the gadget works and then (and only then as it is too often the case) you will also understand what the instructions mean. You have used your intelligence to draw useful conclusions from the minimal cues provided by the insufficient instructions and the gadget itself. Thus, *intelligence is what you use when mechanical rule following does not work.* So, blind rule following is not intelligence and therefore rule-based artificial intelligence is not and will never be real intelligence at all!

Intelligence tests have been used to give a measure of the intelligence of the subject. These tests seek to measure intelligence-related general cognitive abilities that do not require specific knowledge. Therefore it is supposed that the results of the tests are not influenced by training, as learned knowledge is not required. What exactly do the intelligence tests measure? Can we find the mechanisms of intelligence by inspecting these tests? We can try to find that out. Some examples of typical tests are given here.

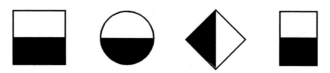

Fig. 5.10. Which one does not belong to the group?

In this first example you are requested to spot the odd one out, the one that does not fit in the group. For this purpose you must find out what is common with the objects and which object does not have the common feature.

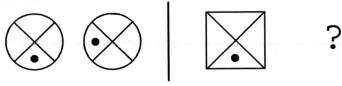

Fig. 5.11. What kind of figure is missing?

In the second example we have two example figures and a third one that is to be succeeded by another one in the style of the example. Now you must recognize the manipulation that yields the second figure of the example pair and apply that manipulation to the third figure.

In the third example we have to figure out the result of paper folding. This calls for the ability to manipulate mental imagery.

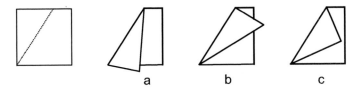

Fig. 5.12. The piece of paper is folded along the dotted line. Which one do you get, a, b or c?

Verbal tests are also used. These may be like the following:

A puppy is to dog like a kitten is to what?

In this case the detection of common relationship is requested.

Normally a large number of individual tasks like these would be given and these would have to be completed in a limited time. Only the brightest subjects could complete all the tasks and in this way the number of correctly completed tasks would reflect the intellectual capacity of the subject.

So to the final intelligence test. We are now supposed to find out what it takes to solve these tasks. It may be possible to find some rules that could be used to solve some individual tests and these rules could even be easily implemented as a computer program. However, this is not the point as I have already excluded intelligence as step-by-step rule following. Instead, we must look for what is essential in general terms in these examples given above. This would constitute the basis of intelligence. It goes without saying that actually we would have to study much larger set of intelligence-demanding tasks than these here in order to have real scientific proof, but these examples will do to illustrate the point.

We can see that the intelligence tests relate to the detection of what is same, what is common, what is different, what is changing, what is missing, what is the relation, what is associated. Also mental manipulations and the ability to imagine different possibilities are involved. Fast completion of the tasks would call for focusing the attention on the relevant and the evocation of proper associations by the slightest cue. These are processes of perception and cognition as discussed before, nothing more. Therefore, in general terms, intelligence relates to perception and cognition and especially would involve the ability to spot cues, often minimal, and to draw useful conclusions from these. This is something that we do all the time, not only at intelligence tests.

Thus, the prerequisites for general intelligence would be fluent perception processes, the availability of concepts and experience, rich associative connections between these, low evocation thresholds and good working memory capacity.

Creativity and Imagination

It has been argued that computers are not intelligent and will never be because they are not creative; creativity is a mental capacity that only humans can posses. Creativity manifests itself in arts; writing, composing, painting, sculpture, etc., and also in engineering; design and inventions. Creativity involves the combination of novel ideas, skill and knowledge. "Inspiration", a kind of "divine influence" has been proposed as the source of the frenzied flow of novel ideas that sometimes seems to accompany creative processes.

What exactly is creativity? The Freudian psychoanalytic view maintains that creativity arises from the sublimation of sexual drives. I agree that sex may enhance creativity, especially in efforts to get some, but to the disappointment of some of you I will not delve into this here no matter how uplifting it might be. Instead some more practical theories are considered here.

Random variation theories propose that creativity arises from random excitations of information and the filtering of the most suitable combinations of these. This theory can be easily implemented as a computer program, but experiments will quickly show that no genius-like results will freely follow. Still, there have been some applications, for example for the generation of new corporate names.

Can we make the random variation method work if we apply evolutionary principles there? We could produce a number of initial schemes by random combination and variation, apply the principle of "the survival of the fittest"; identify then the best alternative and take this as the basis for further random variation while rejecting the rest. This process would then be repeated until the desired outcome were reached. Evolution has produced impressive results so surely evolutionary principles should work here, too?

But does the principle of "the survival of the fittest" really describe evolution? Is the scheme that works best at the moment really the best alternative for further development? Good realization of a bad principle may be better than an initial bad realization of a good principle. Engineers know that many times a new, eventually superior technology produces initially results that are inferior to those produced by the old technology.

The early mechanical television (ca. 1925) of the inventor John Logie Baird did not produce a good picture according to present day standards, but the contemporary all-electronic television faired even worse. When the U.S. inventor Vladimir Zworykin presented his all-electronic television in 1925, the picture was so bad that his superiors tried to prevent him from continuing this seemingly useless pursuit. It took many years of hard work to bring the performance of the all-electronic system to the level of the mechanical television, but then the rest is history.

Likewise, the transistor was born in 1947 as a miserable piece of artifact, a supposed replacement for the radio tube. For many years to come the early transistor did not work well, it was not a good amplifier, it could not amplify high frequencies, it was noisy and was not able to deliver any amounts of power. The radio tube excelled in each of these points and was getting even better. In evolutionary terms the transistor should have died a quick death, the simple "survival of the fittest" rule would have guaranteed this. Electronic television and transistors were kept alive through their early years only by the insight of their stubborn developers.

It is obvious that biological evolution does not rely on "winner-takes-all" rules either, since inferior products also find their corner and are allowed to develop. Evolution does not produce only one result, instead it produces a spectrum of creations whose existence and further development are interrelated to each other in a complex way. Thus the "survival of the fittest" rule in the strict sense of "winner-takes-all" is not really used by evolution.

The development of new inventions and technologies is a creative process, but the "survival of the fittest" rule in the strict sense is not useful here either. Instead we humans can do better than evolution by setting goals and predicting results, having insights about the possible usefulness of presently inferior solutions. We can also learn from mistakes, we can analyze why some schemes did not work and we can fix the problems or at least avoid the repetition of useless experiments. Artificial creativity by "evolutionary" computing schemes that seek to find the one and best result by selecting only the instantaneously best solution for further development while rejecting the rest will most probably fail.

No matter what kind of process is used, creativity is a product of imagination. The better we can imagine something new, the more creative we can be. Imagination involves the evocation of different imaginations, verbal and iconic, their modification and combination. Imagination does not arise from nothingness, the source of our mental imaginations can only be our environment and the associatively interconnected knowledge in our heads. The amount of knowledge and experience is therefore crucial in creativity. Random evocation of imaginations will not lead to intelligent results (modern "art" created in this way is not an exception), therefore the process must be guided and the results evaluated. Thus the creative process would involve the accumulation of knowledge, topically primed evocation of possible imaginations and the evaluation of these. This process is also iterative, the steps are repeated until satisfactory results are achieved.

The phenomenon of sudden insight is not excluded here. Mental concepts are represented by neural signal activities. An idea, a combination of concepts may be developing under the awareness threshold as neural signal pattern combinations and break ups. When the pieces fit, that is when the connective signals become strong enough via the binding of all pertinent relationships, the intensity of the respective neural pattern will exceed the awareness threshold and the idea will suddenly be perceived.

CHAPTER 6

EMOTIONS AND COGNITION

What are Emotions

We all know what emotions are. Emotions make life interesting. Without emotions Hollywood would be out of business in no time at all. Love, hate, jealousy, horror, despair, fear, they all are the stuff that great movies are made of. But movies and novels apart, what good are emotions for human cognition?

Indeed, the rational view has it that emotions have no part in cognition and reasoning, instead they are atavisms from lower stages of evolution. Rational thinking and emotions do not mix and therefore emotions have no role in science or artificial intelligence. One plus one is two whether it makes me happy or not. It is also my personal experience that the laws of physics and engineering are not affected by my emotional states and logic is not subject to emotional interpretation. No wonder then that many books on cognition do not cover emotions at all.

However, an alternative view has it that emotions are an integral and necessary part of cognition. Things and affairs are supposed to have subjective emotional significance and this in turn is seen as a guiding value for quick responses, learning and approximate reasoning. Emotions are also seen as important motivational factors. We do things out of curiosity, out of interest, fear, anger, envy, jealousy, guilt, etc. The study of human motivation would also be the study of emotions.

Some examples of emotions are given in the table 6.1. The inspection of this table should already give us an idea what emotions are for. We could well write a novel by inventing some characters and picking a handful of emotions for them from this table. Emotions seem to give a short-cut to responses, motivation, direction and style for action, what do we need more? Who needs reason as long as we have emotion?

Indeed, if emotions were evolution's lower stage answer to cognitive problems then low-level primitive organisms should be able to cope with emotions only, without any higher intelligence and logic at all. This indeed might be possible as stories about celebrities and politicians seem to prove.

Table 6.1. Some emotions

satisfaction	dissatisfaction
expectation	disappointment
pleasure	displeasure
joy	sorrow
happiness	depression
hopefulness	despair
curiosity	caution
enthusiasm	indifference
boldness	fear
love	hate
like	dislike
ecstasy	anger
sympathy	antipathy
self-confidence	hesitation
astonishment	horror
admiration	contempt
pride	shame
etc.	etc.

Emotions are evoked by triggering events. These events may be external or internal, emotions may be evoked even by one's own thoughts. Emotions are felt, they are accompanied by a subjective experience. Emotions have also physiological symptoms, a body response. Emotions may also cause an expressive reaction. We tend to act out our emotions even though in many cases we might very well prefer not to show them.

Emotions may be accompanied by a variety of physiological symptoms from mild to more severe, either alone or as a combination. For example, excitement and fear make the heart beat faster, sadness slows it and happiness elevates the breathing rate. Common physiological symptoms are given in the table 6.2.

Table 6.2. Emotion-related physiological symptoms

facial expressions	changes in breathing rate
vocal modulation	changes in heart rate
crying	changes in blood pressure
smiling	changes in skin conductivity
laughter	changes in pupil size
blushing	pilo erection (goose bumps)
paleness	stomach symptoms
shivering	loss of bladder and bowel control
sweating	etc.

Facial expressions are the most common physiological symptoms of emotions. When we are sad, we look sad, even without our conscious awareness of the fact. There are different facial expressions for happiness, sadness, surprise, anger, fear and disgust. These expressions seem to be universal, not culture based, and are readily recognized by everyone. Smiling and laughter are the same everywhere.

Vocal modulation is also common. When we are angry we speak up, astonishment may make us stutter. Shock, sadness and despair may make our voice tremble. Excitement makes us speak faster, sadness and depression slower. Usually we can detect and interpret vocal modulation easily without any conscious effort and copy or counteract the projected emotion; if somebody displays anger by yelling at us, we get angry, too. We can detect vocal modulation caused by a smile even over the telephone, without seeing the speaker. Successful politicians, preachers and casanovas know the value of emotional vocal modulation.

The more severe physiological symptoms are usually related to emotions of excitement, fear and horror.

There are many theories of emotions, what they are, how they arise, how they are perceived, how they affect behavior. For our purposes a constructive theory that allows artificial implementation would be needed. In the following I will propose one such theory. Before that a discussion about the emotional significance evaluation mechanisms is needed.

Pain and Pleasure

Pain and pleasure are highly subjective states of mind with reactive consequences. A theory of pain and pleasure must explain what these reactions are and what is their cognitive and physical function. Pain and pleasure are also characterized by their specific feel. Any worthwhile theory of pain and pleasure must also explain this feel, how it arises and how it is perceived, why it really feels like something.

Cognitive neuroscience has been able to find out some basic facts about pain and pleasure, for instance the neural transmission mechanism of pain signals, the role of pleasure centers in the brain, the effect of some chemicals on the perception of pain and pleasure. However, neuroscience has not yet been able to explain what causes the subjective feel of pain and pleasure. This is an unfortunate situation for an engineer, who hopes to design a conscious machine able to feel pain and pleasure, and therefore I will later on propose mechanisms that could explain the subjective feel of pain and pleasure.

When we hurt ourselves we feel pain. Also internal physiological problems can cause pain. Self-inflicted pain causes us to discontinue immediately any activity that is causing the pain so that further damage would be minimized. Pain inflicted by others makes us aggressive and ready to retaliate. In overpowering situations pain may make us submissive. The event that caused us pain will be also instantly imprinted on our mind. Pain caused by injury or illness makes us unable to concentrate on anything else, we must lie down and rest and give ourselves a chance to recuperate. Intense pain may also make the unfortunate sufferer moan and writhe. Thus, pain has the following effects: aggression, submission, discontinuation of the pain-causing activity to avoid further damage; enforcement of rest for recuperation; emotional memory making so that pain-causing acts might be avoided in the future.

In humans pain signals originate from pain sensors, free nerve endings in the skin and elsewhere that react to cell damage. Pain signals are transmitted to the brain via two kinds of fibers, thick ones with myelin insulation and thin others without myelin. The myelinated fibers conduct faster and cause sharp sensation whereas the unmyelinated fibers conduct more slowly and cause dull prolonged burning sensations. These fibers project eventually to the somatosensory cortex[78]. PET scans show that during perceived pain the activated brain regions are the somatosensory cortex, frontal cortex, cingulate cortex and thalamus. If frontal cortex is not activated then pain is not felt.

The somatosensory cortex obviously indicates the position of the injury. The thick fibers projecting here are in close connection to the touch and pressure detecting sensory pathways. Damage to tissue also evokes touch and pressure sensations, therefore the position of the damage point can be resolved by the learned position associations of the touch and pressure sensations even if there were no such association available for the pain signal itself.

The cingulate cortex is an evolutionarily old part of the cortex and has connections to the hippocampus. On the other hand the hippocampus is known to be involved in memory making therefore it can be supposed that pain-related memories were established via this route.

The feel of the pain necessitates the activation of the frontal cortex. However there are no pain sensors in the frontal cortex or elsewhere in the brain, therefore what could account for the feel of the pain? Now here is the real problem. The neural pain signals seem to be similar to other neural signals in the brain. If this were the case, if the pain signals were indeed similar to the other neural signals, like those originating from the eyes or ears, then why would the pain signals be *felt* as painful, why don't the other, similar signals feel like anything? What could then be the mechanism for pain, the cause for the specific *feel* of pain?

The meanings of visual signals are grounded in the outside world, in the seen objects. We have learned to associate the signals with the objects and properties that they represent like shape, color, etc., and this gives the basic meaning of these signals. However, the feel of pain is not grounded in this way to sensed entities *because pain is not a property of a sensed entity.* Pain sensors do not sense pain, the sensed entity is cell damage and the caused signal indicates only that pain is to be evoked. Thus the feel of pain is not a representation, instead it is a system reaction. The pain signals themselves do not carry the feel of pain, instead the feel arises from the effects that these signals have on the system and this in turn depends on the way the signals are connected to the system. The non-representational nature of pain is also obvious from the fact that we cannot memorize the feel of pain and evoke it afterwards. We can remember that we had a headache, but we cannot evoke the headache itself. If pain were a representation then we should be able to evoke it at least to some degree like heard sounds and songs or mental images of seen objects. However, we can only label pain and have a representation for this label.

Pain can be alleviated by blocking the propagation of pain-related signals. This can be achieved by a variety of chemicals, especially those that affect synaptic transmission. It is also known that pain can also be alleviated by focussing attention to other things. The gate-control theory of Melzak and Wall[61] proposes that pain signals may be blocked in the spinal cord by feedback signals from the brain. On the other hand hypnosis may also be used to block pain. Major surgery has been performed without any anesthesia, under hypnosis only. It has been speculated that hypnosis achieves this via chemical means, endorphins released by the brain. However, it has been shown experimentally that this does not seem to be the case[13,95]. Therefore hypnosis, too, would seem to achieve pain blocking via attention diversion. So it seems that there are two methods to alleviate pain; by chemicals that block pain signal transmission and by diverted attention.

Thus pain and attention seem to be closely coupled. Pain "demands attention"; it disrupts any attention that is focused on any ongoing task. Obviously pain signals are transmitted to every modality in the frontal cortex and the message, so to say, is "stop whatever you are doing and try something else so that this signal might stop!". This is because the pain signal itself does not know who should do what to stop the damage and therefore it has to broadcast its message to everybody and thus disrupt the attended processes within each modality. Pain does not allow the other modalities to relax, instead it tries to stop their present activity and start something else. It is exactly this general cross-modality broadcast that makes us moan and writhe when in pain. I consider this disruptive broadcasting as a fundamental property of pain and *I would like to go to as far as to propose that the subjective feeling of pain is indeed caused by attention disruption* especially in the frontal cortex area. Now you, my clever reader, may ask: if this were the case then pain could indeed be alleviated by fixing attention heavily to other things, but on the other hand surely the feel of the pain should also stop altogether if we avoided the attention disruption by giving in and fixing our attention to the pain itself, however this does not seem to be the case? Fair enough, this can be partly tested. There have been a couple of times when I have accidentally hurt my fingertip with a needle. In those cases when I have focused my attention to the feel of the pain (that has not been easy, the brain would not like to do that because initially the feel of the pain increases) the feel of the pain has indeed vanished and only a kind of neutral sensation has remained. I do not suggest that this is a proof, scientific proofs do not come that cheap. I also advise strongly against any experiments like this, do not hurt yourself in order to see how pain works. However, this might be an indication that there could be something there. But how about more general cases of pain, headaches for instance? In those cases the attention cannot be focused on the pain because the pain is not point-like. There are numerous points of origination and as many paths and circuits of pain, each demanding their own attention. Therefore the attention disruption cannot be avoided by focusing attention to one of these pain signals.

Now to more pleasant things. Pleasure is a feeling that we want more. When we are reading a good book the pleasure of reading captivates us and we will ignore all external stimuli. Likewise a bite of chocolate makes us want more, we want to sustain the pleasant taste.

Pleasure has the following effects: Continuation of the pleasure-causing activity to sustain the feel of pleasure; attention focus on the pleasure-causing activity and exclusion of attention to other stimuli; memory making so that pleasure-causing things and acts could be identified and repeated in the future.

Pleasure, like pain is not a property of a sensed entity. There is no pleasure to be sensed and represented, instead pleasure is a system reaction that can be evoked by various sensations. It seems that in the brain there are regions that are responsible for these system reactions and thus can be labeled as pleasure centers. In certain animal experiments an electrode was inserted into the test animal's brain and a switch was provided so that the animal itself could administer an electrical stimulus to the part of brain where the electrode was inserted. When the electrode was connected to the medial forebrain bundle, the test animal would start and keep on stimulating itself as soon as it realized the association between the required act of administration and the effect. The animal would continue to do this until exhausted or the experimenter turned off the electric current. It is assumed that the animal did this because the stimulus generated pleasure[17,101]. The pleasure captivated the animal's attention and did not allow its redirection, leading to sustained action.

I am proposing that pleasure, like pain, is connected to attention, but the attention capturing mechanism of pleasure is different from that of pain. While pain uses brute force to disrupt the attention within modalities pleasure tries to sustain its attention focus by having non-related circuits and modules relax. In this way only the pleasure evoking activity will be continued while other activities are suppressed.

The Beautiful and Ugly, Perceptual Pleasure

Beautiful objects are attractive and evoke pleasure and desire. Ugly objects are repulsive and evoke displeasure. Perceived looks, beauty and ugliness are important emotional criteria that guide our everyday choices be it the selection of clothes, cars, mobile phones or anything else. Without our emotional perception of beauty and ugliness the fashion business would be in great difficulties.

Why, then, are some cars beautiful and others quite ugly? Why are some women merely beautiful and when others are really stunning? Which features make things beautiful and desirable? Are beauty and ugliness inherent properties of things? But then again some can consider an especially ugly car as beautiful while others would disagree. Examples like these would seem to show that beauty and ugliness were only our own subjective percepts.

What makes one face more attractive than another? Judith Langlois and Lori Roggman created pictures of artificial faces with a computer by averaging various numbers of real individual faces[72]. When a number of subjects were asked to rate the attractiveness of these faces it was found out that strongly averaged faces were considered more attractive than most individual faces. As an explanation it was suggested that averaged faces would better represent the category "faces" and would therefore be preferred.

However, when Victor Johnston performed similar experiments and exaggerated facial female features he seemed to notice that these enhanced faces were even more attractive than the averaged faces[52]. Slightly larger eyes, thinner nose and fuller lips made the original average female face look more desirable. Johnston suggested that the perceived beauty of female faces relates to sexual selection process where indicators of fertility would be looked for.

But there are also other objects of beauty. Look at examples of art, cars, architecture, etc. The beauty of these is hardly related to perceived fertility. Therefore sexual selection process can explain only a small part of general beauty perception.

Take the ten most beautiful sports cars. Assemble a new car using the most beautiful parts of these, a fender here, windscreen there, other parts as needed and weld them together. Surely the result should be more beautiful than any of the originals, as this new car combines the most beautiful parts of these. In reality things may not turn out in this way. What we get is a chimera, an ugly mix of forms. Our efforts fail because the new car does not posses uniform style. Here we have one cue to the essence of beauty. The features of a beautiful object must follow a uniform style, they must perceptually fit together.

Earlier I proposed that the perception process combines the effects of actual sensations and inner expectations of these and the resulting match, mismatch and novelty conditions can be indicated by specific signals. The system would strive towards the situation where the expectations match the actual sensations, the cognitive process would be interpreting correctly the sensations. In that situation no need for attention shift would remain. There would be no need to form new binding patterns, this would be a state of minimum effort. I have also argued that pleasure would be related to a state of sustained attention. Sustained sensory match is obviously a state of sustained attention, therefore it should also evoke pleasure. Likewise sustained sensory mismatch condition should evoke feelings of displeasure.

Now I propose that match pleasure is one component of the sensation of beauty. Match pleasure arises from the successful prediction of sensed form. The prediction may arise from memory as conceptual expectation, a kind of average idea of the object, as is the case with averaged faces. When an averaged face is matched against this kind of inner concept of faces, best possible match occurs and maximum match pleasure is evoked.

The sensed object itself is another source of expectations. Good continuation of the features, symmetry and consistent style all suggest a template against which the individual parts of the object are matched.

This kind of good matching and continuity that leads to perceived beauty has also an unexpected connection to engineering. Airplane designers know that "what looks good flies good". An aerodynamically perfect form has good continuity and will therefore evoke match pleasure and will be perceived as beautiful. The same applies to electronic engineering, for instance printed circuit design, especially for high frequency circuits. This phenomenon is also known in architecture and civil engineering, good designs usually look good. However, all good-looking designs are not necessarily good as demonstrated by the notorious Millennium bridge in London.

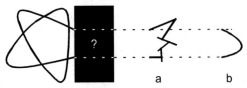

Fig. 6.1. A sensed object creates expectations about its parts. Which one of the figures, a or b would be under the black square? Which alternative would be more pleasing?

Match pleasure generation mechanisms are obviously universal and do not depend on individual preferences. Therefore they do not explain the subjective component of beauty assessment.

Subjectivity may arise via the associated emotional significance of objects. Individual experiences may associate pleasure and displeasure with objects, which then would be perceived as attractive or repulsive. Exaggerated female features could be interpreted as inviting with the promise of pleasure.

Thus, averaged faces are attractive because they evoke maximum match pleasure. However, the emotional component of female faces could be increased by exaggerating female features. The evoked emotional pleasure could well compensate for the slight loss in match pleasure and in this way the manipulated faces could be deemed to be even more attractive. This cannot be guaranteed though due to the subjectivity of the emotional component and some people might experience quite the opposite effect.

The Good, the Bad and the Rotten

Candy is good and sour milk is bad because they taste and smell so. This is already known by little children. Indeed the main purpose of taste or gustation and smell or olfaction is the detection of edible and non-edible substances, to sense which substances are good and which ones are bad and should be avoided.

It is better to smell than taste, because in this way fewer molecules need to be in contact with the organism and the possibility of poisoning can be minimized. Smell can also operate over distances, airborne particles can indicate the presence of food and danger. Odors are also used for chemical communication by animals.

The resolution of the olfactory sense is good; we can detect a large number of different odors ranging from pleasant to revolting. On the other hand the resolution power of the gustation is not very good. There are only five basic tastes: sweet, bitter, salty, sour and umami (the taste of monosodium glutamate). Sweet tastes definitely good, the others may taste good unless they are very strong. The perceived taste of food comes from the combination of taste and olfactory sensations.

The effects of perceived smell and taste are similar. Good smell and taste are pleasant and make us want more. Bad smell and taste turn us away and make us avoid and reject their source. Facial expressions accompany the perception of strong taste and smell. Sweet taste makes us kind of smile, bitter taste like that of a lemon causes quite a different expression. Strong stench may make us retch. These facial expressions would seem to be direct reflexes caused by the chemical nature of the sensed substances. I noted earlier that there are different facial expressions for the emotional states of happiness, sadness, surprise, anger, fear and disgust. Would it be just a coincidence that taste and smell sensations can cause rather similar facial expressions? I don't think so, instead I would say that the "good" and "bad" detected by smell and taste senses are indeed related to emotions in a basic way.

Why should we smile when we feel pleasure? Why should we be pleased when we see others smile to us? What is the connection between the emotional state of pleasure and the facial expression of smile? Sweet taste brings out a certain facial expression, smile, obviously as a chemically caused direct reflex and initially without any emotional content. At the same time sweet taste also causes pleasure, the feeling of wanting more, as a hard-wired reflex.

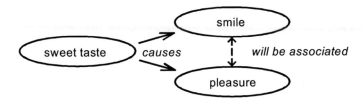

Fig. 6.2. Sweet taste brings out smile and pleasure causing them to be associated

The repeated coincidence of pleasure and smile causes them to be associated together so that smile becomes the facial expression for pleasure. The associative connection works both ways, pleasure may evoke smile and vice versa, smile will have emotional significance. When we see others smile at us, we will imitate them and thus evoke the associated pleasure. This feel of pleasure will thus be associated to perceived smiles and later on the mere seeing of smiling faces will evoke pleasure. In a similar way also the other facial expressions may acquire their emotional significance.

The adjectives "good" and "bad" relate directly to the respective taste and smell sensations but have also more general meanings. Also the basic tastes are used metaphorically to describe entities and states of affairs: "Sweets for my sweet, sugar for my honey" or "Things went sour and I was bitterly disappointed, this whole business stinks". How does this kind of metaphor and generalization arise?

Let's consider the taste "sweet". Sugar is sweet, but we may use the word "sweet" to describe other things and affairs as well. Sweet taste and pleasure are causally and associatively connected. If an entity causes pleasure sweet taste may also be evoked via the associative connection between pleasure and sweetness. Now the original entity and sweetness are perceived simultaneously and may be associated together. In this way "sweetness" acquires an abstract meaning that is no longer directly coupled to a taste sensation.

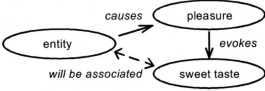

Fig. 6.3. The generalization of "sweetness" concept

A similar process is applicable for the other tastes, too. However, the more general point is that in this way the concepts of "good" and "bad" can be extended to cover a variety of cases. As "sweet" is "good" then also the entity will be "good". In this way events and entities in general may be emotionally labeled as good or bad.

Theories of Emotion

A successful theory of emotion should explain what emotions are and what is their function, how they arise, how they are perceived, how they are represented in the system and why they feel like something. Here we have also another requirement. Emotional processes should be explained in a way that allows artificial realization.

It was stated earlier quite generally that emotions are psychological and physiological reactions to triggering events. These reactions involve also a specific feeling. When we are afraid, we *feel* fear and our heart is pounding. When we are happy we *feel* pleasure and our heart has a steady beat. The common sense view says that first there is a trigger event, then the feeling follows and thereafter the physical response driven by the feeling.

William James felt in 1884 that there is something wrong with this common sense view. When we feel we are perceiving something and what is perceived is the feeling. What do we perceive when we feel fear? Is fear some kind of a perceivable substance within our brains? Obviously James did not think so. Instead he proposed that the trigger event is directly followed by the physiological reactions and it is the perception of these reactions that constitute the specific feeling. This idea, nowadays known as the James-Lange theory[49,69], has been misunderstood and ridiculed ever since: "Do we feel fear because we are running away or do we run away because we feel fear?" Yet, there is experimental evidence that there is some truth in here. In the Schacter and Singer experiment volunteers were injected with a drug that produced physiological arousal symptoms. Half of the subjects were informed about the effects of the drug, the others were not. During the experiment the uninformed subjects misinterpreted the effects of the drug as emotional feelings caused by the situation[70].

The James-Lange theory can be criticized by the point that many emotions seem to have quite similar physiological reactions. If this is the case, then what causes the perception of different feelings? Also, physiological reactions should take some time to build up, yet we seem to have the emotional feeling instantly at the onset of the triggering event. The Cannon-Bard theory[50,69] tries to avoid these objections by stating that the feeling and the accompanying physiological reactions are independent of each other even though they may occur together. The Cannon-Bard theory may indeed solve the build-up time problem but then it also loses the explanation of the feelings, it does not explain what is actually perceived when feelings are felt, as feelings here are supposed to be independent of perceivable physiological reactions.

Some early theories of emotion proposed that there is a small number of basic emotions and all the other emotions are only blended combinations of these. For instance Robert Plutchik proposed that there are eight basic emotions, namely anger, anticipation, joy, acceptance, fear, surprise, sadness, and disgust. All the other emotions would be derived from combinations of these[83]. Each of these basic emotions is supposed to be directly related to survival enhancing patterns of behavior.

Later theories of emotion have included the aspect of cognitive interpretation. These theories like the "two factor theory"[69] propose that emotions involve the interaction of the perception of the triggering event, physiological reaction, cognitive interpretation and feeling.

Modern neuroscience seems to support the newer theories of emotion. The neural pathways involved in emotional processes seem to consist of the pathways from sensors via the thalamus to cortex, hippocampus and amygdala. Various interconnections and feedback loops are also involved here. The cortex is supposed to be responsible of the cognitive interpretation, the hippocampus is supposed to be involved in memory making and the amygdala connects to the brain stem and is thus responsible for the physiological reactions[51]. Thus those theories of emotion that integrate the processes of perception, cognition, memories and physiological reactions would seem to have some neural basis and support.

The System Reaction Theory of Emotions

The designer of an emotional machine needs a constructive theory of emotions that outlines a practical realization. If some kind of combinatory theory is to be tried then the basic elements to be combined should be chosen so that they are realizable within the framework of the artificial system. Thus certain emotions cannot be taken as elementary unless these are defined as realizable system processes.

Therefore, for the purposes of cognitive machine design, I outline here another theory of emotions, one that may not necessarily be accurate in neurological terms but yet contains essential similarities to biological emotions. Thus I propose that emotions could be considered as interactive combinations of perceivable basic system reactions with eventual cognitive evaluation. These emotions would thus also determine the style of the system's responses.

By cognitive evaluation I mean here the associative connection to multisensory perception processes and memories as well as match/mismatch detection. Basic system reactions would be direct pre-wired responses to certain elementary sensations. These elementary sensations would originate from the elementary senses; taste and smell senses, pain and pleasure senses and match/mismatch/novelty detectors. I consider these senses elementary because they offer simple information that suggests clear specific responses as opposed to the visual and auditory sensations that need complicated processing to determine what is where and even then may not suggest specific responses.

The basic system reactions would be simple, rather automatic hard-wired stimulus–response style responses to the elementary sensations, needing no cognitive evaluation of the situation. For instance a good response to a bad taste or smell would be the rejection and withdrawal from their source in order to avoid possible poisoning. Likewise the source of good taste and smell could be accepted and approached. Pain causing action should be discontinued, pain causing adversaries should be aggressively fought. These are simple responses that enhance the prospects of survival. The proposed elementary sensations and related system reactions are given in the table 6.3.

Table 6.3. Elementary sensations and related system reactions

Elementary sensation	Related system reaction
Good; taste, smell	Accept, approach
Bad; taste, smell	Reject, withdraw
Pain; self-inflicted	Withdraw, discontinue
Pain; due to ext. agent	Aggression (retaliation)
Pain; overpowering	Submission
Pleasure	Sustain, approach
Match	Sustain attention
Mismatch	Refocus attention
Novelty	Focus attention

These system reactions are useful as such and a simple organism could well do with these without any further cognitive processes. However, it would be even more useful if a system reaction and the related situation, the cause for the reaction, could be associated and memorized. Thereafter this association would be available as experience. For instance bad taste could be associated to the visually perceived representation of certain substances so that later on there would be no need to taste them again and subject oneself to poisoning. Pain could be associated with dangerous adversaries so that they could be avoided in the future. Pleasure could be associated with certain activities so that these could be exercised when possible. This association process would also allow the generalization of the stimulus; sweet taste, goodness, could be associated to generally pleasant objects and situations. Bitter or disgusting taste, badness, could be associated to generally bad situations in the way discussed before. In this way the basic system reactions would be available for more general external world and imagined situations, ones that are no longer directly related to percepts from the elementary sensors.

It should be noted that generally system reactions are not necessarily the same as a system's actual responses. The stimulus–response style of operation is applicable only for trivial cases. In general cases system reactions prepare the way and give direction for actual responses. They set the positive or negative framework, the "mood" for actual responses, which are also determined by context and the system's general state, needs, intentions and experience.

In order to perceive and internally represent system reactions the system needs system state sensors and their related perception process. The internal representation of system reactions allows their associative connection to other sensory modality percepts and emotional memory making. These connections are shown in the model of fig. 6.4.

According to this model pain, pleasure, taste and smell (good/bad) sensations as well as match/mismatch/novelty states are able to initiate the basic system reactions such as those listed before and also other physiological reactions. These system reactions are perceived by system sensors and their respective perception process. The system reaction percepts are associatively connected to visual, auditory, tactile, etc., percepts and vice versa, visual, auditory, etc., and percepts are associated with system reaction percept representations. The system reaction representations themselves are able to evoke some of the system reactions, at least to some degree.

Thus a visual or auditory percept of an entity that has been associated with, say, pain, will act as a triggering event and will evoke the representation of pain and this in turn will evoke some pain-related reactions. Feedback via the cross-connection path back to the visual memory process will elevate the signal intensity of the original pain-evoking percept. This in turn will ensure that the pain-evoking visual percept will become and remain as the prime object of visual attention. The cross-connection to auditory area facilitates inner verbal reports or comments that may affect, accompany, amplify or subdue the perceived system reactions.

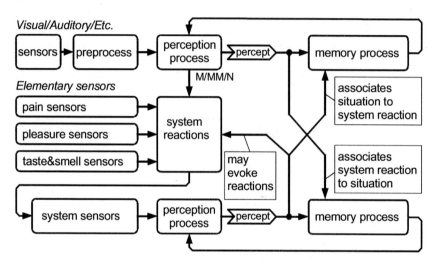

Fig. 6.4. The connection of system reactions to perception and cognitive processes

The memory process of this model is also the source of thoughts, inner speech and inner imagery. These are brought back to percepts via the corresponding feedback loops. System reaction percepts and system reactions may also be associatively evoked by these thoughts or memories with associated system reaction percepts without any actual sensory stimuli. This is consistent with common experience, thinking of sad memories may make us sad, thinking of income tax may make us angry.

Instead of a single system reaction there may be several reactions going on simultaneously, also conflicting ones. These combinations determine the system's response and behavior in the given situation. I am proposing here that these system reaction combinations are the phenomena that we call emotions. In the table 6.4 some combinations of basic system reactions are listed and the proposed corresponding emotion named.

In table 6.4 "pain", "pleasure", "good", "bad" and "novelty" designate their system reactions as described before. Arousal is not indicated in this table, as it is a basic reaction common to almost every emotion.

Table 6.4. Emotional states as combinations of basic system reactions

emotion	\multicolumn{9}{l}{system reactions of:}
	pain	pleasure	good	bad	novelty	aggression	submission	approach	withdraw
curiosity					+			+	
astonishment					+			+	+
caution					+				+
fear	+			+					+
anger				+		+			
desire		+	+					+	
love	+	+	+					+	
happiness		+	+						
sadness				+			+		
disgust				+					+
envy	+		+	+		+	+		
horror				+	+				+
despair	+			+			+		+
resignation	+			+			+		

According to this table "curiosity" is in addition to arousal the combination of the system reactions of "novelty" and "attraction". This makes sense; curiosity means being attracted by novel, not familiar things. "Astonishment" is here a combination of the system reactions of "arousal" and "novelty", "attraction" and "withdrawal". Astonishment is more than surprise. An unexpected gift or tax refund is a surprise but astonishment here is a state where the system is not able to interpret a sudden novel situation properly and thus counteracting attraction and withdrawal reactions may both arise. "Fear" is a kind of anticipation of pain and is therefore a combination of internally evoked representation of pain, the system reactions of "bad", "arousal" and "withdrawal". "Desire" is here a combination of the system reactions of "arousal", "pleasure", "good" and "attraction". This again depicts reasonably well the actual state of desire. Likewise the other emotions in this table are described as such combinations of system reactions that produce the functional state and response of the specific emotion.

However, this is a simplified table for the illustration of the basic idea only and no claims about the psychological accuracy are made at this point. The system reactions are more varied and graded than depicted here. There is also the temporal dimension that is not captured in this table. For instance the emotion "astonishment" involves the sudden onset of the unexpected, unexplained novel percept.

Likewise "disappointment" would involve the expectation of something good that does not materialize at the expected time.

The good point of this approach is that emotions are here defined by the combination of system reactions. These in turn are entities that can be defined constructively as I have already done. Therefore emotions defined like this can be synthesized mechanically in systems that have the respective system reactions.

Emotions *feel* like something. Can we produce the feel of emotions in this way? I have proposed that "pain", "pleasure", "good" and "bad" get their specific feel via the attention affecting mechanisms as described before. "Aggression" and "submission" have also similar system level effects. Attention affecting mechanisms can be realized artificially, thus effects that seem to be related to the feel of emotions can be included in the mechanical realization of emotions.

Emotional Significance

Our perceptual and cognitive processes produce a continuous stream of percepts and thoughts and these are subject to emotional evaluation. Perceived entities are rated as beautiful, ugly, attractive, repulsive, good, bad, pleasant or unpleasant, painful, threatening or having neutral emotional significance. These emotional ratings are used to control our attention and they also may trigger full-blown emotions, which in turn suggest the line and style of responses.

These emotional ratings are also tagged to our memories. We will learn to long for action that produced pleasure, "this was good, I'll do it again". We will learn to avoid events that turned out bad and unpleasant, "I will never ever go back to that restaurant". Without emotional tagging every memory would have the same neutral significance and we could not benefit from our experience.

Emotional significance affects also learning and memory, it helps the brain to distinguish events that are worth memorizing. Our old memories usually depict some significant or unusual moments of our life. We have happy memories, we have sad memories, we remember embarrassing moments. It seems that a memory is more easily recalled if the present emotional state is the same as that of the memory. When we are sad we tend to remember sad things.

Higher level habituation and sensitization may also be explained in terms of emotional significance. Initially a new stimulus has emotional significance due to its novelty, it must be assessed so that necessary actions could be taken if any. In habituation the emotional significance of a repeated stimulus decreases as the stimulus is being assessed harmless and requiring no reaction. On the other hand, if the repeated stimulus cannot be ignored due to its intensity or other property sensitization may occur. In this case its emotional significance increases and the subject will be more and more disturbed by the stimulus.

Motivation, Needs, Drives, Goals

What makes us do something? We do things because we have to; we have to eat, drink and sleep. We do also things that bring us pleasure and we do things in order to avoid displeasure. We do things out of curiosity and anger. Things like these motivate us to do something.

Motivation is the reason to do something; motivation is here understood as the process that initiates and directs a system's actions towards goals. Motivation has been divided into internal and external factors. Early theories described internal factors as innate instincts that were triggered by environmental stimuli. Newer models describe internal factors as drives; psychological states that arise in response to internal physiological states, needs like hunger, thirst, etc. Once an internal need is detected a drive sets in and directs or pushes the action towards the satisfaction of the need. On the other hand incentives are identified as the main motivational external factors. The expectation of a reward pulls the action towards the given goal. Obviously fear or expectation of a punishment would seem to have the opposite effect.

Fig. 6.5. The push–pull metaphor of motivation

External motivational factors may be learned. In practice internal and external factors may interact in many ways. For instance an internal need cannot be satisfied unless certain external conditions are already met; we cannot eat if we do not get food first. More complicated concepts of motivation have been proposed like achievement motivation; a supposed internal need to seek success in one's life[71]. Also emotions are an important motivational factor, needs arise from the pursuit of pleasure, curiosity, revenge, etc.

The list of our active needs, drives and tasks to be done can be compared to a stack with priority order. What to do now, what to do next, the resumption of a task that was discontinued a while ago are primed by our own needs and cues; perceived opportunities from the environment. We all know the weight of uncompleted tasks, they are lurking in the stack and distracting our attention. We also know the satisfaction of completed tasks, the pleasure from results matching our expectations.

Motivation seems to arise from internal needs, external incentives and conditions and momentary emotional states in a complex way. These motivational processes would seem to call for specific structures in the brain responsible for motivational and rewarding factors.

Indeed, in the human brain the hypothalamus has been identified as the main structure that is related to the needs of feeding, drinking and temperature control, also known as the homeostatic motives[16,100]. Another part of the brain, amygdala, seems to be responsible of emotional responses. On the other hand the pleasure centers in the brain are supposed to form a reward system that integrates feelings of reward for the satisfaction of all the basic needs[17,101].

Humor and Laughter

We are proud of our sense of humor, a supposed hallmark of humanity and something that machines can never have. But is that so?

We like to laugh because laughing feels good. However, we cannot laugh at will, we cannot make ourselves laugh real laughter on command. Laughing is characterized by intermitted breathing, related sounds and pleasing relaxation. This relaxation can even go as far as to make a person roll on the floor. Laughter, like a hiccup, is a result of certain kind of system state and in order to induce laughing we must produce that system state. This system state can be produced directly by tickling but since tickling is not usually socially acceptable we have invented indirect ways to produce the required system state, that is via jokes, humor and comedy. We tell jokes and watch silly sitcoms in the hope of easy laughter.

Why do jokes make us laugh? Little babies may laugh at sudden unexpected phenomena if they feel themselves safe. Bopeep games, a bouncing ball never seen before, a sudden noise or any unexpected thing may make a baby laugh. These same things may also make a baby cry if fear is evoked instead. Clearly there is no humor there, instead the elements involved are astonishment and relief. Astonishment involves the sudden experience of the unexpected and unexplained, relief involves the match pleasure that follows the successful interpretation of the event and the return to familiarity. This course of events provides the jolt that makes the system oscillate, laugh, for a while.

Now to the jokes. Why is it so that we must first *understand* the joke before we can laugh at it? Why do familiar jokes not make us laugh, even though we surely understand them? What does this understanding involve? "One fine day a man went to work and in the afternoon he was quite tired." This is a very short story that you all understand, but did it make you laugh? I don't think so, because it did not provide a jolt, it was not a joke. A joke is a story that evokes the elements of astonishment and relief. A joke is usually built around double meanings. The story is developed along one meaning until the culmination point or punch line where the other meaning should become suddenly evident. To understand a joke means that the double meaning has been captured. The sudden realization that everything in the story does not fit and the subsequent perception of another interpretation that makes things fit again gives the kick. When we hear the joke again, the element of astonishment is no longer there and laughter will not be induced. There is a clear mechanism here and there is nothing that could not be implemented by a suitable associative machine.

Why do some people consider dirty jokes as the best ones? Why should certain inappropriate things be funny? It seems that the inappropriate nature of certain things is already able to provide a jolt via the realization that something has been said that should not be said. This can be applied to conventional jokes; an amplified effect can be achieved if the punch line reveals an inappropriate double meaning. However, some people may find these jokes emotionally so offensive that no laughter is evoked.

But then not all laughing is related to jokes. Social laughing seems to be induced in groups by remarks, gestures and cues that do not really constitute a proper textbook joke. Still, the basic mechanism of laughing is there and I would propose that in this case laughing is induced indirectly, as an association to pleasant situation. Also the mirroring effect is there (a trick that also the pre-laughed silly sitcoms try to utilize). You laugh because others laugh and this again makes the others laugh even more. Positive feedback is operating there and it is known that when a system is on the verge of oscillation only a small kick is needed to trigger the action. Thus no elegant humor or fancy jokes are needed or actually used. It is strange though that afterwards the same bunch of people considers themselves to have an exceptionally good sense of humor because they were laughing so much. ·

Music Perception and Emotions

Why should we like to listen to music? After all instrumental music is physically only a special case of noise that does not actually represent anything. We do not really need music for survival and could very well do without it. Therefore it is difficult to see any evolutionary pressure that would have shaped our brains for musical perception, yet all human societies, old and new seem to have music. The perception of music is strongly emotional and emotions, as I have argued before, are related to cognition. Therefore any cognitive theory that claims to explain the processes of the mind must also solve the problem of music; how the perception and emotional effects of music are related to cognition. The explanation of the mind is not complete if a part of human mentality, music, remains outside it and any proposed cognitive processes should be seriously doubted if they cannot accommodate music. Basically a theory of music must explain why rhythms, melodies and chords may sound pleasant, why this can take place even when a piece of music is heard for the first time. It must also explain how and why music evokes feelings and affects our moods, how music evokes emotional states, like the Dr. No theme from the James Bond movies evoking suspense and excitement, Jobim's Bossa Nova melodies portraying sadness and happiness at the same time and Sibelius' Valse Triste being gloomy beyond belief. The theory must also explain the motor effects of music, why certain rhythms make us tap our feet and feel like dancing.

When discussing the connection between music and cognition the first question should be: Is there universal music independent of cognition? I dare to answer swiftly: No universal music exists, "music" is only a categorical name that we have given to artificially produced sound patterns that we find perceptually interesting. I also reject the hypothesis of special brain circuits for music perception and processing as evolutionarily implausible. Our brains have not evolved to accommodate music, instead we have developed music to suit our perception processes. Therefore I shall propose the following. First, music perception utilizes general pleasure inducing auditory perception mechanisms and cognitive processes. Secondly, some of the emotional effects of music arise from associations with emotionally significant events and memories. Thirdly, music, especially musical rhythms, is related to the body's own rhythms and rhythmic motions via mirroring processes.

First I will consider how the auditory perception process could explain the sense of basic elements of music; rhythm, chords and melody and their perceived pleasantness or unpleasantness.

Those of us who have experimented with audio signals may have noticed that pure sine tones with limited duration sound quite neutral while continuous sine tones are irritating. Notes produced by musical instruments are not pure sine tones, instead they contain harmonic components and sound definitely more pleasant than pure sine tones. As most natural sounds contain harmonics it is obvious that the brain learns to associate fundamental frequencies and their harmonics, therefore the fundamental frequency would predict the harmonics. This prediction would be successful with instrumental notes and would lead to perceptual match pleasure in a similar way that I have described in the "Beautiful and Ugly" section.

When two notes with almost the same fundamental frequencies are played together, dissonance takes place, the notes sound unpleasant. Earlier I have proposed that pain and its lesser form displeasure are related to attention disturbance. In this case attention cannot be focussed unequivocally to one or the other note as they are too close to each other to be resolved properly but not close enough to be fused. Therefore attention hunts irregularly between the two notes and displeasure follows. Thus dissonance displeasure would be direct result of the design of the ear and the auditory perception process and as such not tuned to the perception of music, the situation rather being the other way around.

Chords are pleasant combinations of three or more notes played together. According to the above propositions chords should sound pleasant if dissonance is avoided and even better so if some harmonics can be fused. This is indeed the case in practice. In fact musical scales are designed to minimize dissonance and to allow harmonic relationships.

Music is physically heard as one note or simultaneous notes at a time, but the impact of music comes from the succession of notes, the melody, rhythm and tempo. The perception of music relies on what has been heard just a moment ago and what has been heard over a span of time, from the beginning of the musical piece. Based on this history the perception process makes predictions about what is coming next. Forthcoming notes can be predicted by the direction of notes, ascending notes predict a higher note and descending notes predict a lower note. Melodic patterns predict the change of direction of notes. Rhythmic patterns predict timing. Successful prediction generates match pleasure. There may be several alternative predictions available, if the major prediction fails then an alternative prediction may still provide match pleasure. In this case transient interest enhancing novelty and then a surprise match is also experienced, a jolt akin to that of humor. Examples of these effects are key changes and syncopated rhythms.

Dissonance and prediction match/mismatch pleasure and displeasure explain only a part of the emotional effects of music. Therefore causes other than perception process mechanisms must also be considered.

The very purpose of auditory perception is to evoke responses and many of these are emotional. Sudden noises in the dark are a warning and call for attention. The thunder of lightning, roar of a lion, the howl of wolves, the hiss of a snake may all evoke fear. The sounds of laughing, crying, moaning, wailing can directly evoke emotions via mirroring mechanisms. It seems quite plausible that the reproduction of the rhythmic pattern and tonal quality of these sounds by musical instruments could evoke respective emotions at least to some degree. The eerie sounds of the Theremin instrument may paint a scary picture of werewolves or tortured souls, the slow tempo of a funeral march may remind of the deep slow breaths of moaning with all the associated emotion.

A piece of music is an auditory percept and as such is able to accumulate associations with other percepts, be they visual, emotional or other. A piece of music may be associated with a certain situation and may later on evoke memories of that situation. Strong emotional significance alone helps memory making, therefore emotional situations are more easily associated with music and thus memories evoked by music may have an emotional content. We like to listen to music that evokes pleasant memories or perhaps the emotional content of the memory only.

Motor sequences such as walking, running, tapping the feet, etc., involve rhythmic motor command patterns. The execution of motor sequences do not always take place in complete silence, footsteps can also be heard, therefore auditory percepts of motor rhythms are occasionally available. Thus two-way associative connections between certain auditory rhythms and motor command patterns can be learned. Musical rhythms with timing that imitates the timing of certain motor sequences like marching, running, or dancing may therefore induce the execution of these actions.

On the other hand fast paced motor rhythms are associated with excitement and slow motor rhythms to tranquil states. The complex connections between emotional states and physiological symptoms as discussed before could thus explain the exciting and tranquilizing effects of music. When we are sad and tranquil, we like to listen to music with slow tempo that matches our slow motor rhythms, then again when we are happy we find faster paced music more agreeable. It would seem that match pleasure and mismatch displeasure mechanisms are operating here, too.

There are connections between music and language. Syllables, words and sentences have their own timing and intonation patterns that are detected separately and are used to help the recognition process. Emotional states modulate motor acts, their intensity and pace. This applies as well to speech, therefore the vocal modulation, that is the rhythm and intonation of speech carry also emotional content. If a piece of music is designed to imitate the rhythm and intonation of oral narration, perhaps an emotional one, then obviously the general form and emotion of the narration will be perceived; the piece of music tells a story albeit without words and actual concrete meaning. The emotional content of the narration, encoded in the rhythm, pace and intonation is carried over by the piece of music.

There is no mystery in music. We have created music because we have found the various effects of artificial sound patterns pleasant. A machine with human-like perceptual, emotional and associative processes could also enjoy and create music.

CHAPTER 7

LANGUAGE AND THOUGHT

What is Language?

We all know what language is, don't we? Because of common language I am able to communicate my ideas to you, the respected reader of this book.

But, language is not only for communication. Even more importantly language is a thinking tool. We would need language even if we were the only person on earth. We use language in the form of silent inner speech to label, comment and ponder our moment to moment situations. Our ideas seem to clarify themselves and get ordered when they get a linguistic expression. Little children talk aloud all the time, not to others but to themselves, and those of us who have had the rather tiring opportunity to observe this may have noticed that this speech is often in the nature of a commentary. Language seems to be essential to our thinking, so much so that some people equate thinking, consciousness and inner speech.

The basic questions about language would be: How do we learn words and how do we acquire grammar, the rules that say how words are to be assembled into sentences so that a given idea can be expressed? How do we understand words, sentences and stories? How is language connected to thinking processes? These are important questions if we are ever to build a thinking machine that could learn a natural language and use it in the same way as humans do.

The linguistic approach assumes that a language consists of vocabulary and grammar. Therefore the understanding of a language calls for the acquisition of the lexicon that gives the meanings of the words and the grammar that gives the rules how sentences are to be parsed. Indeed, vocabulary and grammar are what you acquire when you learn a foreign language. You memorize grammatical rules and words–create the vocabulary.

Initially you will translate back and forth between your native language and the foreign language and you will also consciously apply the grammatical rules. Eventually if you become fluent, you will begin to "think" in the foreign language; to produce sentences and understand without back and forth translation. Also you will forget the grammatical rules yet you will continue to produce grammatically correct sentences. The loss of the grammatical rules and the non-existence of translation become apparent when you are asked to actually translate from this foreign language or to describe its grammar.

So it seems that the initial approach of translation and the use of grammar are only temporary devices used to facilitate learning. Most decidedly we do not learn our native language in this way!

This linguistic view is also called the horizontal approach because the basic assumption of the sufficiency of vocabulary and grammar alone lead to the study of the relationships between words only. Therefore in this approach the meanings of the words can be ultimately defined by their relationship to other words. For instance the meaning of the word "cat" could be "a furry and soft animal". More accurate descriptions could be easily devised if necessary. In this way the meaning of every word could be defined without any help from the outside world. Thus the language in itself is enough, syntactical relations between words and sentences carry all necessary information for deduction, reasoning, etc.

However, this approach can be criticized for the lack of ultimate grounding of meaning. The definition of the meaning of the words will be eventually circular like "large means big", so what does "big" mean? "Big means large." It really does not help even if we had more and more words involved in a definition that eventually folds back to itself. This is a familiar problem, for instance Philip Johnson-Laird has protested that theorists have ignored the relation between language and the world. "Semantic networks perpetrate the 'symbolic fallacy' that meaning is merely a matter of relating one set of verbal symbols to another"[46].

The alternative, vertical approach to language assumes that the meanings of the words are derived from entities in the outside world or, more precisely, from our percepts of these entities; our inner imagery. Therefore the meanings of words are established by their connections to their respective outside entities. The outside entities have their own relationships and the language and grammar only reflect these. Therefore the language is a description of inner imagery that is evoked by sensations. Thus, thinking, deduction and reasoning involve the manipulation of inner imagery and the verbal description of the result.

However, we seem to have the everyday experience that perhaps most of the time we do not have inner imagery at all. Also we seem to be able to reason without inner imagery. For instance, if A is bigger than C and C is bigger than B then which is smaller, A or B? We can solve this without any imagery of A, B or C, or even without knowing what they actually are, without any grounding in external world entities.

The vertical approach can also be criticized for the difficulty, if not impossibility of the grounding the meanings for abstract words.

Yet, we engineers do use inner imagery in creative thinking. We have ways to represent information graphically as drawings and circuit diagrams. Many of us can visualize complex circuits in our minds and visually reason how they work without the use of a single word. So what is the truth? The truth may be that both vertical and horizontal approaches are needed. The meanings of concrete words must be grounded in the real world, but on the other hand the linguistic apparatus must be able to work by itself by the rules that it has extracted from the real world.

Language, Thought and Inner Speech

We all have this silent or inner speech inside our heads and if asked, many of us would say at first that thinking is in fact this inner speech. This may not be exactly so, but inner speech is nevertheless an important thinking tool. I would compare inner speech to a short hand notation, one that allows us to track and bring to awareness contents of our mind in a condensed way so that complex mental representations and their relationships can be more easily mastered. Inner speech also enables us to remember our thoughts so that these can be used as the basis of further thoughts and can be compared to each other. However, inner speech is only one facet of thinking processes. The interaction between inner speech, inner imagery, feelings and sensations should not be ignored. We can excite ourselves emotionally by our own verbal thoughts only and vice versa, our emotional state affects the way we think.

There are two main linguistic theories of thought, the natural language theory and Jerry Fodor's mentalese theory[33]. The natural language theory of thought states that thinking is performed by the natural language that we use in our inner speech. The mentalese theory of thought states that thinking is performed by a subconscious language, the "mentalese". The natural language inner speech that we have would only be a translation of mentalese. We think we know how our natural language is learned but how do we learn our mentalese? Fodor maintains that mentalese is innate, everybody is supposed to be born with it, its vocabulary and grammar. I may be mistaken here, but it seems to me that since no mentalese concepts are learned then there would have to be ready-made innate concepts also for, say, transistors and quantum dots and people would have had these concepts already in the Middle Ages even though they did not have natural language words for them. I find this hard to believe.

How exactly are thoughts, the inner speech sentences generated? If "thinking" involved nothing more than inner speech then the inner speech itself would have to generate new sentences by the power of the vocabulary and syntax. We are aware of the inner speech, therefore we should be aware in advance of the next thought that we are going to have, or at least we should be able to see how it is being assembled if there were nothing else to it. However, this is not the case, we are not aware how our thoughts arise. Therefore "thinking" must involve processes that we are not aware of. Also, our thoughts are about something, they refer to our needs, moods, outside world, etc., there is a gist, a general idea. Our thoughts are not random albeit syntactically correct strings of words.

The natural language theory would seem to have difficulties here in explaining how thoughts arise in the first place while the mentalese theory would have an answer to that; the natural language inner speech is a translation from the subconscious mentalese. But, the solution offered by the mentalese is only a trick, the original problem is only transferred to mentalese; how do the sentences of mentalese arise in the first place?

A working theory of thought and inner speech must answer the question how thoughts, the sentences of inner speech are generated in the first place, how their meaning is grounded and understood and how inner speech is related to inner imagery, feelings and sensations. I consider both the natural language theory of thought and the mentalese theory insufficient in these respects and will therefore propose later on another approach.

Grammar and Syntax

Many students of a foreign language feel that grammar is a nuisance, an unnecessary pain in the neck. So why to study grammar? We do not think of grammar when we speak in our native language, and in fact we may not even be able to explain the grammatical rules that we seem to utilize. More paradoxically, the better we master a foreign language the less we need to consciously consider its grammar. So what is the point of learning grammar in the first place? Why do we need grammar at all? Why don't we express ourselves just by strings of words without any grammar and syntactic rules? What is the additional value of syntax that is supposed to make it indispensable? Okay, let's assemble a couple of sentences as strings of words without any syntax and see what we get: "Tom hit Jack", "Bill go home must school from direct", "this yours is". The first sentence seems clear enough, so we do not need syntax here? But, without syntax can you say who hit whom? Did Tom hit Jack or did Jack do the punching? The second example is messier and we can only guess the meaning. Does Bill have to go to school directly from home or vice versa? In the third example we cannot deduce whether this is a question "Is this yours" or a statement "This is yours".

Now we can see the purpose of syntax. The mere string of words does not convey unique meaning. Sentences are like stories, they describe states of affairs like who does what to whom, when where, etc. In order to understand a sentence we need to know the relationships between the words. Syntax is used just to indicate that, syntax can be seen as a convention how to arrange words so that doers, owners and objects, etc. can be identified. Therefore syntax is background information that is needed in the correct interpretation of sentences.

Syntactic relationships can be expressed by the proper order of words as in the English language or by inflection, by specific word endings as in the Finnish language so that the word order does not matter. Some relationships can be expressed by prepositions like "to", "from", "under", "over", etc. These prepositions can be used in inflected languages, too, in addition to the inflection. Now at this point it somehow does not seem to go without saying that due to the flexibility of the inflection mechanism the expressions in inflected languages, at least in Finnish, seem to be far more accurate than in the non-inflected languages.

Syntax provides also the means to express and distinguish present, past and future action, conditional cases, commands and questions, etc. Syntax brings redundancy and predictability, too and in this way makes the understanding of speech easier.

The grammar of a language includes syntax, phonology or sound patterning and semantics or meaning. Here a distinction can be made between the linguistic "text book" grammar and the real "grammar" actually used by the brain, these are most probably not the same thing. Accordingly my aim here is not a machine that implements textbook grammars but a machine that can learn and produce language, dare I say, in a "more natural way".

How is the grammar learned then? Learning by trial and error with teacher feedback can be proposed. However, an almost infinite number of invalid sentences can be constructed, therefore the number of required trials might be prohibitively large. Noam Chomsky argued that the rules of any grammar are so complex that they cannot possibly be learned and yet these rules are possessed by every adult, therefore they must be known already at birth[20]. Babies can grow up to use any language and any grammar, so at birth they would have to posses all the different grammars of the world. It is difficult to see how this could be possible, therefore Chomsky proposed that in the infant brain there exists one innate universal grammar only, which during growing-up is transformed into the grammar of the used language. This may well solve the problem of grammar learning, but at a cost and this is not accepted by all. For instance Terrence Deacon laments: "Universal Grammar is a cure that is more drastic than the disease. It makes sweeping assumptions about brains and evolution that are no less credible than the claim that children are super-intelligent learners"[27]. Indeed, where would this built-in Universal Grammar, so complex that it can cover every natural language, have come from? Deacon proposes that grammar can be learned after all via trial and error learning. In practice children do not sample the full range of word orders, etc. Instead they are strongly biased in their choices. Where does this bias, this restrictive information come from? Deacon proposes that it comes from the evolution of the language itself. Languages must be learnable at each stage of their evolution in order to survive, only learnable changes can be passed on. Thus the grammar of a language is necessarily such that it can be learned[28].

It is nowadays fashionable to call evolution to the rescue whenever a problem like this is encountered. However, there may be simpler explanations available. What information is available to a young child when he tries to generate a sentence? Firstly there is the idea, the state of affairs to be verbalized. Then there are the vocabulary and sample sentences. Now the point is, are a few sample sentences sufficient for the generalization of the convention — the grammar — that says how the sentence is to be constructed? Why would children be biased towards correct syntax if they didn't have grammar? Would it be that the poor imaginative power and undeveloped structures of young brains really do not allow the evocation of all possible alternatives so that even the slightest cues from example sentences would be enough?

This is an important question to the designer of thinking machines. If Chomsky has it right, then the only way to build a linguistically thinking machine is to preprogram the grammar of the used language. However, if we could build a machine that can learn language and grammar on its own then Chomsky's idea would be wrong and that would be the end to one theory in linguistics.

The Multimodal Model of Language

It is obvious that neither the horizontal nor vertical view of language alone is sufficient for the purposes of thinking machines. It also seems that the natural language theory of thought and the mentalese theory are insufficient for the explanation of inner speech and its relation to mental imagery and sensations. Also the idea of innate grammar may be suspect. Therefore another approach is attempted here, a view that combines the horizontal and vertical approaches and connects them to perceptual processes and furthermore assumes no innate grammar. I call this approach the *multimodal model of language.*

The multimodal model of language begins with the assumption that there exist a representation and memory plane or a space for each sensory modality. These planes learn and acquire their representations via the perception process. Each plane contains representations of its own kind; the visual plane can handle visual objects, the auditory plane can handle sound entities, the touch plane tactile entities, etc. Representations within a plane can be bound into groups or assemblies and these in turn can be linked to each other. These representations can also be vertically connected to representations in the other planes.

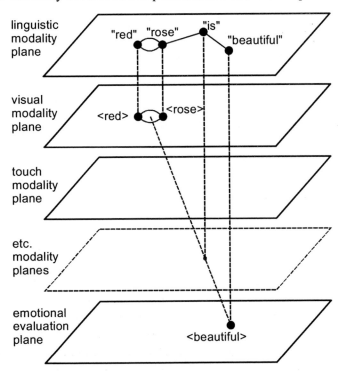

Fig. 7.1. The multimodal model of language

The basic meanings of the representations are causally grounded in the external world entities and sensed bodily conditions. The plane for spoken language resides in the auditory modality. Therefore the meanings of its representations, words, are grounded in heard sound patterns. But here is the twist. Due to the assumed associative connections words may also be associated with other representations in the same or other modalities. Now a word no longer depicts only its sound pattern, in addition its representation in the linguistic plane will represent another thing, another sensory percept.

The diagram of fig. 7.1 tries to illustrate the multimodal model of language, its connections and relationships between the words within the linguistic modality and the vertical grounding connections to the other modality planes. The representations within each modality plane may have horizontal connections to other representations within the same plane and also vertical connections to other modality planes. The diagram is not to be taken as a static construction, instead the connections between representations should be seen changing over time. The temporal relationships of the representations can also be represented and these representations again can be connected to other representations, both horizontally and vertically.

As an example a sentence "red rose is beautiful" is represented on the linguistic plane by the connected representations for each word. The meanings of the words are vertically grounded to the visual plane representations of <red> and <rose> and to the emotional plane representation for <beautiful>. Here the entities <red> and <rose> are percepts that relate to external world entities. <Beautiful> is a percept that relates to a system state as <beautiful> is not an actual property of an external object, instead it is a judgement made by the sensing subject. However, the actual message and meaning of the sentence "red rose is beautiful" does not emerge from the individual words only. The message is neither "red = beautiful" nor "rose = beautiful", instead it is meant that a rose that is red is beautiful. Therefore the entities "red" and "rose" must be bound and used as an assembly to convey the intended meaning.

The binding of entities like <red rose> in this example is supposed to take place independently on each plane, also on the linguistic plane. Therefore the linguistic plane alone would be able to store the idea "<red rose> is <beautiful>" without actual grounding of the meanings of the words. Thus the linguistic plane would be able to answer questions like "what is beautiful"–"red rose" and "what is red rose"–"beautiful" by the power of connections between bound entities only and without any reference to the actual meanings. Here the binding of <red> and <rose> on the visual plane take place due to the spatial coincidence, both properties are detected at the same position. The idea "<red rose> is <beautiful>" corresponds to the shifting inner attention from the percept <red rose> to the simultaneous percept of <beautiful> and the storage of the link. The binding and linking are conveyed to the linguistic plane and after a few examples like this a pattern emerges, the linguistic plane will bind and link the words of heard sentences in a similar way even if no grounding to other planes exist. This is the basic mechanism for syntax acquisition. *Thus syntax is seen here to arise from the real world relationships between entities*; the syntax reflects these. As I see it, the real world relationships are globally more or less the same, the grammars of natural languages must have syntactical structures to represent these and this is the reason behind the functional equivalence of these structures, not any proposed innate universal grammar.

Words can be learned as sound patterns without any reference to entities in other modalities. These words would be connected to other words and sentences only, therefore their meaning would remain abstract. Thus certain linguistic concepts may not have any counterpart in other modality planes at all. Then again, it is possible that there is no linguistic expression for every possible representation in the other modality planes. Little children are sometimes victims to a similar effect, you can see how they struggle for words for an idea that they so clearly seem to have. This state of affairs is also familiar to those of us who need to talk in a foreign language.

Each modality plane has a built-in memory capacity. However, this memory is not like a computer memory with addressable memory locations. Within a computer a distinction is made between the memory addresses and the actual data that is stored in the memory locations; data can be retrieved by the activation of the intended memory address. Here no addressable memory locations exist and no addresses are used. Instead information is stored associatively so that given representations can evoke other related representations. These associations are learned in the ways that were outlined earlier, including the usage of evaluated emotional significance.

It can be seen that this associative mechanism of the multimodal model of language provides an additional and very effective way to store and retrieve information; the linguistic way. Without a linguistic plane information could be stored only as associations of direct visual and other sensory modality representations, that is, sensations that represent directly the sensed entities. The linguistic plane enables the compression of these entities and their relationships by substituting these by verbal labels. In this way language functions like a shorthand notation for the storage of the represented ideas.

However, this storage does not usually take place in the form of rote memorization of sentences, instead nets of associative links are assumed. We cannot easily recite by heart even short passages of text that we have just read, but we can tell what the text was about.

In the case of spoken language the linguistic modality is connected to the auditory modality. However, this is not the only possibility, a language need not to be based on auditory tokens, instead any other sensory modality will do as long as distinguishable tokens can be conceived and perceived. We can have sign languages, tap languages, etc., each with their own "words" and grammar.

According to the multimodal model of language real world entities, actions and relationships are represented by vertical connections between the linguistic plane and other modality planes and the horizontal connections directly on the linguistic plane and indirectly on the other planes. Thus language is seen as a description of the external word and its events as well as the system's inner states; needs, motives, emotions. However this description need not to be direct. Metaphoric representation is possible and may arise via common properties between entities.

Words and Their Meanings

Little children learn their first words because we teach them. We repeat words so that the child can imitate them. We point to things and label them. We also name actions: Come, give, take, eat. First the child will learn to understand these words, he will be able to show the objects when the word is given. Later on he will be able to pronounce these words and use them for communication. At first words are used alone then the child will learn to combine them into two word sentences and much later the ability to produce unlimited length grammatical sentences appears. A child will also invent his own words for things, these words are usually some kind of derivations of words that the child already knows. When a child possesses a vocabulary of about 100 words (I have counted up to 40) new words with meanings that can be easily pointed out become scarce and parents obviously begin to think that the child already understands everything. Thereafter they no longer actively point out and name objects and from now on the child must capture new words and their meanings on his own.

There are two kinds of words, concrete and abstract. Concrete words have a meaning that can be directly perceived and indicated, like a seen objects or action, sound, taste, smell, bodily sensations, movements, etc. Concrete words may be taught by giving the word and pinpointing the entity to be named. This method is also called "ostensive definition"[21].

In the multimodal model of language the meaning of a concrete word is usually linked vertically to a representation on one of the modality planes. Thus an ostensive definition establishes a two-way associative connection between a linguistic plane entity, that is, a word and a respective entity in one of the other planes. Either one of these associatively connected entities can evoke the other. For instance, the word "cat" can evoke an inner representation of a cat in the visual plane and vice versa, a visual plane representation of a cat can evoke the word "cat". Therefore an entity that is associatively connected to the respective word via an ostensive definition is called here the basic meaning of the word.

However, the meaning of a word is not restricted to only one association. A word may also evoke everything that is linked to the representation of its object as secondary associations. For instance the word "cat" may evoke an inner representation of a cat and this in turn may evoke ideas like "cats are cute", "angry cats may scratch", etc., not necessarily in verbal form, based on the subject's experience on cats. These ideas in turn are used to determine the required response; to ignore or not, to approach or avoid, to be pleased or not, etc. Therefore via the association between the word "cat" and the inner representation of a cat these secondary associations will be available as further meanings for the word "cat". Shortcut associative links may be formed between the word and the secondary associations so that the evocation of the original first order inner representation is not necessarily needed. Context and background determine which secondary associations are relevant at any time.

Here I define an abstract word as a word that cannot be taught by ostensive definition. Abstract words by this definition would include words like "abstract", "truth", "economy", "existence", "like", "if", "or", "else", "but", "although", "however", etc. Some abstract words can be quite tricky, as an exercise the reader is asked to define the meaning of the word "but".

The sad truth is, most words are more or less abstract, they do not have a meaning that could be simply pointed out. The easy answer to this problem is: the meanings of abstract words are defined by other words. However, if we were asked to define an abstract word, even a common one, most probably we would produce several awkward sentences towards an explanation and still leave something to be desired. Yet we use these words fluently and nonchalantly when we speak. We also pretend to understand what has been said without trying to figure out the meanings of the words.

Abstract words cannot be learned by ostension. How then is it possible for abstract words to exist in the first place? The multimodal model of language mechanism allows the learning of words as sound patterns without any reference to entities in other modalities. If this kind of non-grounded word is repeatedly encountered, it will become correlatively associated to both verbal and other repeating entities around it. The word will thus become associated to a situation. Many words that nowadays have only an abstract meaning were concrete words initially. How does this kind of abstraction take place? Now suppose that we take a concrete word and don't allow the learning of its direct meaning. What we get is a non-grounded word with connections only to the repeating situations. This kind of abstraction process has been observed in the development of sign language among deaf children to whom no sign language is taught. Older children developed a set of gestures with concrete meanings. Younger deaf children were not able to learn these concrete meanings because the original contiguity between the gesture and the entity was no longer displayed. Therefore they could only develop generalized meanings for these gestures[87]. Similar processes can be observed in the meanings of some new present day idioms.

Linguistic Understanding

Suppose that you are a professor of literature and you ask your students to read a short poem as homework. How do you now test whether your students have actually read it? That is simple. If the poem is short you may quite well ask the students to recite it by heart. But what if you ask the students to read a book and you want to see if they have understood it? In this case you would not ask the students to recite it word by word, as this would only demonstrate excellent memorization. Instead you would ask questions concerning what the book is about. You would ask questions like who did what to whom, when, where, why. Would these questions show that the story is understood? Not really. It is possible to answer these questions quite mechanically by reproducing the relevant parts of the story. Deeper levels of understanding can be tested by asking the student to paraphrase the story or parts of it, that is, to describe it in their own words. Also the student could be asked to give reasons for the situation in the story, to describe what would happen in slightly different circumstances and to predict what could happen next. Furthermore the student could be requested to evaluate the significance of the story.

What happens when we read a book and understand what is depicted there? Do we parse the text, do we search for noun phrases, verbal phrases, etc.? Context-free rule-based parsing has been proposed and used by Artificial Intelligence to find out who in the text is doing what to whom, etc.[107]. However, how do we parse sentences like "Visiting relatives can be boring" and "The shooting of the housewives was terrible"? Are visiting relatives boring or is visiting boring? Who was shot? Sentences like these should make it clear that context-free parsing will not necessarily lead to a correct interpretation of the sentence. Context, the information in adjacent sentences and in the actual situation, cannot be overlooked. For instance if we are talking about a ladies event at a shooting club and note that "The shooting of the housewives was terrible" then there will hopefully be no confusion. What is needed for the correct understanding of language is therefore the ability to utilize context and background information; common sense.

When we read a book we obviously memorize something, but what would that be? We do not memorize stories as strings of words. Normally we cannot recite even short passages of text by heart. However we must memorize something otherwise afterwards we could not possibly describe the gist of a story. Now certain things must be memorized, connections must be established between memorized entities and background information. I am proposing here that reading a story with understanding is not really different from the perception, understanding and memorization of actual events. Certain things happen in temporal order, you will perceive these and have your own subjective interpretations with emotional significance evaluations. You will also make episodic memories of these. Afterwards you will be able to recall and describe what you think happened. A similar process is taking place when we read a book, this time however the perceived entities and their relationships are evoked by the written story. The text of a story evokes multimodality representations and associations as if we were living the depicted event. Some of these are memorized and associative links are made between them and our background information according to their significance.

Thus, story reading is not a simple process of parsing the text, instead it is an active process of perception and memory making. If a cognitive system is not able to understand what happens in the real world then it will not be able to truly understand linguistically represented stories either.

This requirement is satisfied here; the multimodal model of language assumes a cognitive system where real world understanding and linguistic understanding are performed by the same processes.

Language and Communication

People need to communicate with each other. Our society depends on communication. Language allows us effectively coordinate action by questions, answers, comments, requests, commands and warnings. Language allows us to share and store ideas and wisdom. Small talk with no specific purpose gives us pleasure. With language we can flatter, impress, persuade, threaten, intimidate, hurt and humiliate. We can affect and change the attitudes, behavior, beliefs and emotional states of others.

The traditional view sees communication as the transmission of a message via a transmission medium:

MESSAGE → ENCODING → TRANSMISSION → DECODING → MESSAGE

A message is first encoded into suitable symbols that can be transmitted over the medium, then the received symbols are decoded back into the original message. Communication has been successful if the received message is the exact copy of the transmitted message. Information theory (first presented in 1948 by Claude Shannon, 1916–2001) allows us to compute exactly what it takes to encode and decode a message for errorless transmission through a given imperfect medium so that the received sequence of symbols may emerge as the exact copy of the transmitted sequence of symbols. The meaning of these symbols is of no consequence and is not dealt by the mathematical information theory. This is how our mobile phones are made to work and this is how we can receive pictures from Mars. However, this technically highly successful model of communication is too simple for our purposes.

Human communication is not only about sending a message that can be recovered by the receiver, I would even say that *the message is not the message*, instead the message is the intended effect on the receiving party. We do not talk so that the others may hear and recognize our words, instead we speak in order to make others do something, change their minds, get wise. Communication has been successful only if the desired effect is evoked.

Thus the technical communication model is to be modified to involve the intended effect of the message. Communication has now the following phases: Encoding the intended message into verbal, written or other suitable expression, transmission by speech or other media, reception and decoding; interpretation and understanding. Finally, the received communication contributes to an effect on the receiving party. Two-way communication, conversation, involves these steps in both directions and thus involves continuous feedback.

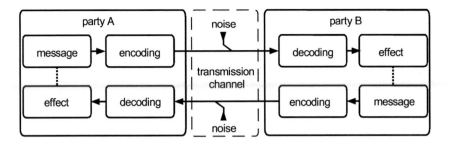

Fig. 7.2. The elements of two-way communication

In fig. 7.2 the party A desires an effect on the party B. Let's say for instance that party A wishes a door to be opened. Therefore party A formulates and encodes the respective message, like "open the door please". Party B receives and decodes the message and if the communication is successful, opens the door.

The transmission channel may affect the communication. Conversation under noisy conditions can be difficult and misunderstanding may result. Normally we may speak so many words per minute, but under noisy conditions we may have to speak more slowly and repeat many times what we say so that the communication speed is lowered. In this example the message "open the door please" could be corrupted by noise and another message might be perceived, like "clean the floor please". Communication has failed due to ambient noise.

The success of communication does not only depend on the ambient conditions like the transmission channel noise. Even more important is the compatibility of the message and the cognitive stance of the receiver. Therefore we must also study the processes of message and effect generation, their emotional and cognitive underpinnings. These aspects are depicted in fig. 7.3.

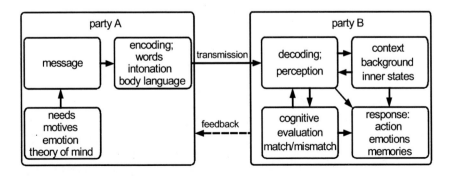

Fig. 7.3. The complexities of human communication

In fig. 7.3 Party A wishes a certain response from the receiver, party B. Party A has certain needs, motives and emotions behind these wishes. Now the message must be formulated so that it will evoke the desired effect and response. Therefore party A should know something about the receiving party B. First of all, common language and common concepts must be assumed. Then the context and background information that is available to the receiver must be considered. A message that is detached from context and implies background information that is not available to the receiver will not be understood.

The receiver's response does not depend on the message directly, instead the message will be cognitively and emotionally evaluated and matched against the receiver's inner states. The receiving party may understand the message, perhaps a request, but still decline to comply with it. Therefore party A should have some idea about the mental state of the party B; party A should utilize the "theory of mind" that will be described later on in this book. Party A should seek to ensure that the mental state of party B would be receptive to the communication. This can be achieved by inducing pleasure, attraction and curiosity. Polite forms of speech have been devised for this purpose. Auxiliary cues may also be used; intonation, tone of the voice, gestures, facial expressions. Mirror neuron action can be utilized here to make the receiver copy displayed emotional states.

Match pleasure and mismatch displeasure are also related to human communication. We are pleased to hear opinions that match those of ours. This is also the idea of "small talk". Find an insignificant every day matter to talk about, then agree with your conversation companion and you will be friends in no time at all.

On the other hand opinions that oppose ours may easily evoke displeasure and aggression. Therefore, if we want to introduce novel ideas, we had better to begin with familiar concepts and proceed slowly so that the mismatch displeasure would be minimized. An occasional reference to accepted concepts would also help as the evoked match pleasure would help to compensate the unavoidable mismatch displeasure.

CHAPTER 8

CONSCIOUSNESS

The Mystery of Consciousness

Humans are conscious. We know what we are doing and why as opposed to being mere sleepwalkers. We are aware of our thoughts, emotions and feelings. We are able to introspect and retrospect. We are aware of our own body and our environment. We can feel pain and pleasure. We are aware of time, the fleeting moment of here and now. We have remembered pasts and expected futures, what we did a while ago, what we are about to do, how these things are related. We are aware of our own existence. We are conscious when we are awake, very much less conscious when asleep, hopefully not conscious under anesthesia, totally and irreversibly without consciousness when dead. When we are conscious we are able to exercise our free will to control our actions and thoughts. Consciousness involves reportability across neural modalities; e.g. if we consciously see an object, we are able to grasp it if it is close enough, we can verbally report what we consciously see and feel. Lowered consciousness can be experienced when tired, just before falling asleep or when intoxicated. We know when we are conscious, we are aware of our own consciousness. It is generally thought that consciousness is the key element of higher human cognition and in fact consciousness would be what separates us from animals and machines.

What is this consciousness then? What happens in the brain when we have a conscious thought? When we are lifting a heavy load we can feel our muscles getting tense, we can feel our heart pounding faster, we may sweat; we can immediately see material processes taking place. But, when we think, we cannot perceive any related material processes taking place, not at least in the brain. Yet, our thoughts can materialize in physical actions. We can think of doing something and physically do that in the next moment.

Everyday observations like these may easily lead to the conclusion that two aspects are involved here, the immaterial consciousness and thinking process and then the material body and external world. French philosopher René Descartes (1596–1650) proposed that this indeed is the case. Mind and matter would be two separate and different kinds of substances. An immaterial self or soul would be responsible for conscious thought and the material body would be responsible for any physical actions. This view is known as *dualism*, the coexistence of an immaterial conscious self and a material body.

The dualistic answer to the problem of consciousness is: Consciousness is a property of an immaterial soul. As the word "soul" is not considered very scientific nowadays the word "self" is often used instead. Dualistic theories usually involve the concept of "Cartesian theater" a place in the brain that displays the products of our senses to the observer self. The self exercises free will and reason and controls our body via the brain.

Dualistic theories are not without difficult problems. First there is the problem of interaction, known also as the *mind–body problem*: How does the immaterial self interact with the material body? Immaterial ghosts (if they existed) may wander through material walls and we are not able to grab them, likewise it would be as difficult for them to cause any material consequences. The mind–body problem manifests itself also in the observed effects of matter on consciousness. If the immaterial soul is indeed conscious independently of the material brain then why is the soul not conscious while the material brain sleeps and why it is possible to induce unconsciousness by blocking the operation of the material brain chemically as in anesthesia?

Descartes proposed that mind and body interact via the pineal gland in the brain — ultimately through divine intervention. It has also been proposed that no mind–body interaction really takes place, mental and physical events that seem to be related only happen simultaneously as they are synchronized so.

The second major and serious problem of dualism is the problem of causal explanation: Humans are conscious because we have conscious souls and that explains that. But how then does the immaterial soul or self produce consciousness, free will and reason, how do we explain that? We have no way of knowing and with material instruments we cannot find out. We cannot even verify the existence of the soul in a material way. Only those hypotheses that can be tested are scientific. We cannot test the existence of the immaterial soul, we have no way to observe its workings, we cannot test any claims about it, therefore the "study" of immaterial soul or self produces only fairy tales and has nothing to do with science.

However, dualism is not dead. It seems that we are easily tempted to accept a little bit of dualism here and there. The modern computer may still be taken as a vague model for the brain. The computer has two main constituents; hardware and software. Hardware is definitely matter, but how about software? How much does a computer weigh without and with installed software? The answer is: exactly the same. So, installed software weights nothing, is software then immaterial and may we equate it to immaterial mind? But then, how can immaterial software determine the operation of the material hardware? In reality software is not immaterial after all, it requires material carrier with suitable distinguishable material states for the representation of information.

The dualistic mind–body interaction problem may, in principle, be solved by denying one of the two substances of dualism, either the material side or the immaterial side so that no interaction is needed. The denial of the material side leads to *idealism*, a theory that denies the existence of all matter. On the other hand the denial of the immaterial substance leads to *materialism*, a theory that states that everything can be explained by the existence of matter and material interactions.

Surely we have proof of the existence of matter? We can see, touch and feel material objects. Who in his right mind would deny their existence then? However, we know that the brain represents external world by neural signals. These signals are not the same as the actual depicted objects and could be generated without the presence of any real objects at all. Our mind has only ideas about things out there; these ideas do not necessarily represent the outer world correctly. Movies are one example of this; ideas of external world objects and events are generated by light intensity variations on a screen. It is solely due to our deficient perception process that movies can evoke illusions of real world objects in the brain. But, we could take this one step further. Illusions of the external world could arise by internal causes alone if suitable neural activity were evoked. We would get the "doorbell effect" at its worst and what we see and perceive would only be an illusion of an external world, which in reality would not exist. Therefore, in theory at least, the non-existence of the external world might be a possibility.

Bishop George Berkeley (1685–1752) went even further. According to Berkeley matter does not exist; there is no material brain either, our percepts of matter are only illusions in our immaterial minds. These illusions are created for us by God. This view surely solved the mind–body problem for once and all, yet it did not appeal to everybody. For instance, Dr. Samuel Johnson (1709–1784) wondered in his letter to

Berkeley what might be the purpose of the sensory organs that we seem to have. What would be the function of the eye, if there were nothing to be seen as all of our visual percepts were implanted into our mind by God? The existence of a coherent material world that can be perceived via senses seemed more appropriate to Johnson. He dismissed Berkeley's idealism concretely by kicking a stone; "I refute it thus!"

Modern science and engineering also reject idealism; matter and interactions that can be studied are assumed. However, principles of idealism have gained a new lease of life in computer games and virtual reality.

Materialism solves the mind–body problem by denying the existence of immaterial substance. However, proper materialism does not deny the existence of mind, it denies only the mind as an immaterial substance. The materialistic answer to the problem of consciousness is: Consciousness is produced by the material brain and the mechanisms of consciousness will be eventually found by studying the brain and its workings. The apparent problem of materialism comes from everyday observations. Thinking and thoughts do not seem to be material as we cannot perceive any material processes taking place when we think. We cannot perceive any "material wheels of cognition" turning in our heads, nor do we perceive the biologically more realistic firing of neurons. However, nowadays there exist a multitude of technologies all able to detect material brain processes that are directly related to thinking. These material processes seem to be the prerequisite to our thought processes and our subjective feelings of say, pain and pleasure. The question is: Are these processes in themselves the same as the actual thoughts that can be reported simultaneously? Why should we perceive our thoughts as such and not as the neural firings that actually take place?

Thus it may be concluded that *any worthwhile material theory of consciousness must explain why our thinking and feelings appear to be immaterial and how these seemingly immaterial processes are related to the material processes of the brain.*

This problem may superficially look like the original dualistic mind–body problem, but it is not the same. In materialism there is no real mind–body interaction problem as no immaterial mind-substance is assumed. Instead, this problem is a perceptual one; why do we perceive our mental processes, which in reality have solid material basis, as immaterial. If we could explain this, then we could also determine whether similar illusions of immaterial mentality could arise in a machine. This works also the other way around. If we could explain the problem in machine terms, then, according to the principles of materialism, we could apply the explanation to the brain also. In the following discussion the materialistic mind–body problem is referred as the *mind–body effect*.

The concept of *emergence* is often proposed here. When neurons are assembled into a large enough system, like in the brain, "higher phenomena" like consciousness, may emerge. The emergent phenomenon is supposed to be something more than the sum of its constituents. In my opinion the concept of emergence may not necessarily be the best of explanations, maybe not an explanation at all, because it would seem to involve a statement of the style: "In big systems it is suddenly like this and there is nothing more to it". As an explanation this is more like a leap of faith.

Another explanation to the mind–body effect will be proposed later on in this book in chapter 18, the section "Ghost in the machine".

The Easy and Hard Problems of Consciousness

Some present day philosophers reckon that consciousness involves both cognitive processes such as those discussed before and also something else that would be responsible for the phenomenal first person feeling of consciousness. Obviously the cognitive processes as such may eventually be emulated by suitable computers and circuits and therefore may be labeled as easy problems, even though so far their practical implementation has not been so easy at all. On the other hand the phenomenal part of consciousness seems to be more difficult to understand and define, therefore it may be labeled as a hard problem.

David J. Chalmers has listed the easy problems of consciousness as follows: "the ability to discriminate, categorize and react to environmental stimuli; the integration of information by a cognitive system; the reportability of mental states; the ability of a system to access its own internal states; the focus of attention; the deliberate control of behaviour; the difference between wakefulness and sleep"[19]. Indeed, these "easy" processes have not yet been very easy to implement, but one can envision ways to handle these with existing technology.

The hard problem of consciousness is related to phenomena that are described by the following terms: qualia or the feel of percepts, phenomenal consciousness, conscious experience, what does it mean when a conscious subject is phenomenally feeling something.

Artificial cognitive machines may be devised to perceive stimuli and to produce proper responses to these and even volition may be emulated in this way. In fact it just might be possible to emulate human cognition without resorting to any kind of subjective experience;

the machine would react to events that cause pain and pleasure in humans with proper responses so that an outside observer might believe that pain or pleasure had been really felt. However, in reality the machine would operate without any felt sensations in a kind of zombie like manner. Would this kind of system, if complex enough, appear to have consciousness? Maybe, and it might even "insist" that it has subjective experiences. If this was the case then how about us? Would it even be possible that our own subjective experiences are not real, that the phenomenal *feel* of qualia is only an artifact of our cognitive processes. This may be a possibility, for instance Dennett has argued along these lines that qualia do not really exist, they are only an illusion within (a philosopher's) mind[30].

On the other hand it might also be possible that there are real *felt* qualia and they have a crucial role in the cognitive processes and therefore human cognition cannot be successfully emulated without the inclusion of any subjective feel. While cognitive functions as such may be more or less easily emulated, so far there have been few if any practical cues about how an artificial system could be made to subjectively *feel* and experience anything, truly feel pain or pleasure for instance. Would qualia-like phenomena arise as a result of the information processing style alone or is something else needed? This is the essence of the hard problem from the engineering point of view.

Intentionality and Qualia

In the philosophy of the mind intentionality means that a thing can represent another thing without being that thing itself. Thus ink stains on paper or electric signals within a computer can represent something that they themselves are not, words or pictures, etc. In reality they are only what they are — ink stains and electric signals. Some philosophers declare that true intentionality can only appear in conscious minds. Ink stains are only ink stains for themselves and for the paper, they do not refer to anything else. For instance Roman Catholic priest Franz Brentano (1838–1917) maintained that referring to something as an object is a mental act, not a physical phenomenon[14]. Thus only conscious human minds could give meaning to the ink stains and computer signals, only humans could interpret them. Further, we alone could give meaning to the mental representations inside our heads, whether these refer to external real world entities or to imagined, even non-existent ones (like the unicorn). Therefore a physical machine could not possess intentionality, its inner signals could not be about anything for the machine.

Moreover, our mental experiences, inner representations are understood to have qualitative properties, qualia, like the blueness of blue, the taste of wine, the timbre of a sound, etc. Qualia are related to intentionality; our mental experiences both are about something and they have the qualities of that something. These properties of our mental experiences are also subjective; they constitute our own phenomenal first-person feeling of consciousness that is not accessible as such to others. The philosophical question about qualia is: what are they and why do they exist?

This I call the weak problem of qualia: How can things be represented by something else and how can these representations be differentiated from each other; why red and green look different, why different tones sound different, why different odors smell different, why the sensation of red is different from the sensation of a sound, etc. Neuroscience maintains that in the brain information is represented by neural signals and signal patterns. Some philosophers are keen to point out correctly that no neural signal is red, no neural signal smells like something, no neural signal pattern sounds like a tone. But then they come to the conclusion: Therefore neural signals and patterns cannot carry qualia, therefore consciousness cannot be mere neural activity. In my opinion these conclusions are logically and technically false.

Let's consider a color TV as an example. Inside a color TV set and color TV studio equipment color information is carried by three signals, one signal for each of the primary colors red, green and blue. These signals are electrical ones and therefore obviously do not have any color. When we observe them by an oscilloscope we will notice that they are exactly similar in general appearance and when seen out of context no inference about the carried property can be made. What then from the TV set's internal point of view could make one signal carry the property of "redness" and the others carry "greenness" and "blueness"?

This "problem" is exactly the weak problem of qualia; how neural signals, each superficially similar to every other, could possibly be about something and carry properties like "redness", "sourness" or what ever. This question is a trivial one and its mystification is strange to any electronics engineer.

Certain signals can depict for instance redness simply because their causal origin is "redness", that is, reflected light's respective spectral properties in outside world. The signals themselves do not have and need not have the property of "redness". All it needs is for the system to be so configured that this causal point-of-origination meaning is preserved. This is not a very weighty restriction and very complicated signal processing operations can still be performed. If you proved otherwise, your mobile phone, video and color TV would no longer work and autopilots could no longer fly planes from London to New York for example. I am not proposing that such equipment is conscious, I am only illustrating the point that signals can be about something for their system and they can carry the properties of that something.

The strong problem of qualia is as I see it: Why some percepts really do feel like something, especially like pain and pleasure and to some extent beauty and ugliness, too. Here the grounding of meaning via the preservation of a point-of-origination path does not really work, because the feel of pain or pleasure is not a property of the sensed entity. Instead these signals only indicate that pain or pleasure is to be evoked. Thus pain and pleasure are not things that are represented, instead they are a system reaction along the lines that I have already outlined. Likewise the feelings of beauty and ugliness are not properties of sensed objects, instead they are subjective judgements accompanied by system reactions.

Now you may say that it is all right to maintain that strong qualia are *related* to system reactions and especially the effects on attention, but why would system reactions *feel* like something to the system? I am arguing here that they do, but to know for sure elaborate investigations would be needed. Perhaps, if we were eventually able to build an artificial conscious system and experiment with it, then this question might become easier to understand.

This question of strong qualia is a tricky one due to its subjective nature and any explanation can and should be suspected strongly. However, there is also the danger of demanding overexplanation. A notorious example of overexplanation is the microwave oven fallacy: "Microwaves heat food because they make the food molecules move and this causes friction and that causes heat". An explanation can be seen as a bridge from the strange to the familiar and to a layman this explanation might seem to be a robust bridge indeed. However, the motion of molecules *is* already heat and the explanation should stop here. Likewise, from the system's inner point of view, attention affecting reactions may be the strong qualia and if this is the case then the explanation should stop there.

Now back to the basic question: What are qualia and why do we have them? I propose that weak qualia, the perception of different sensations as different, are merely the perception of the different signals and signal patterns. Strong qualia, the feeling of something, are temporal behavior patterns of inner attention caused by system reactions. We have weak qualia because we must separate sensations from each other; these arise directly from the style of information representation. We have strong qualia because we have system reactions that affect attention.

Consciousness, Perception, Attention and Memory

Consciousness is about the ability to know and feel one's environment, body and mental states. Sensory perception is the cognitive process that is responsible for information acquisition about our environment and body as discussed earlier.

How about our mental states? These we can access via introspection. But, what is this introspection? We do not seem to have the power to introspect and perceive the states of neuron groups or individual neurons for that matter, yet thinking and thoughts are supposed to be represented by the activity of these. Then how can we perceive our own thoughts? What kind of trick is needed here?

Is it mere coincidence that we hear our thoughts as inner speech? Hearing is perceiving and inner speech allows us to perceive our thoughts as silent speech, obviously with the same machinery that is used to perceive external speech. If we had some means to sense our thoughts as neural firing patterns then we would not have to transform them into heard speech. Similar argument goes for inner imagery, too. In fact it has been experimentally verified that imagined visual imagery elicits activity in the same cortical regions as visual perception[10]. This would show that inner imagery enters consciousness via the same perception process as sensed external imagery. These tricks could allow the brain to operate and perceive the attached meanings of neural signal patterns without any need to be aware of the actual neural activity or its location within the brain.

Therefore perception would seem to be the key mechanism for consciousness and percepts would seem to be the content of consciousness at any moment. Philosopher David Hume (1711–1776) already noted: "I never can catch myself at any time without a perception and never can observe anything but the perception..."[34]. When nothing is perceived, not even one's own thoughts, no feelings, no pain or pleasure, there is no consciousness, therefore perception processes are a necessary prerequisite for consciousness.

Possible mechanisms of attention have been discussed earlier. Now it is time to ask: What is the relation of attention to consciousness?

In many theories it is implied that consciousness is accompanied by attention, or that attention is the very mechanism that determines the content of consciousness. Indeed, by definition attention guides perception and if percepts are the contents of consciousness as argued above then attention does select the contents of consciousness. However, what has now been said is more like the definition of attention and does little to explain either attention or consciousness. What is the connection between consciousness and memory? It is clear that many cognitive functions require short-term memory. The words of a sentence must be simultaneously available for a while so that the correct associations can be made. Mental arithmetic cannot be performed without the temporary memorization of the numbers and results. What one is going to do next usually depends on what has already been done, thus a short-term memory is needed here. Retrospection is not possible without memory.

However, is a short-term memory needed for the phenomenal aspect of consciousness? Can there be any kind of consciousness without a short-term memory? The idea that without memory there is not much consciousness is quite old. Charles Richet, a Frenchman wrote in 1884: "To suffer for only a hundredth of a second is not to suffer at all; for my part I would readily agree to undergo pain, however acute and intense it might be, provided it should last only a hundredth of a second, and leave after it neither reverberation nor recall"[22]. Is that really so? Could we not accumulate these kinds of extremely short "non-conscious" events back-to-back so that their total duration might be hours? After the ordeal we would come out clean and pronounce: It did not hurt! However, was the experience still unconscious? Would you volunteer for that kind of experiment?

Let's consider another example. Did I lock the car doors before I walked away? Sometimes I realize that I do not have any recollection of that. When I go back I see that the doors are locked. Did I do that unconsciously and without any attention? I must have exercised some kind of attention, otherwise I would not have been even able to insert the key into the lock. I just did not memorize the event, was it therefore an unconscious act? Does this very same act, executed by the same motor neurons and motions, become conscious if it is going to be remembered later on? If so, how long should the memory of an event last in order to allow us to call that a conscious act? Or, would the actual act, neural activity leading to the locking of the doors, be the same in both cases the only difference being that the original act was or was not reported to such neural modules that are able to reconstruct it later on as a remembered event? In that case memory itself is not the crucial factor, it is the communication or lack of it between neural modules that makes the difference.

Why are some actions remembered afterwards and therefore deemed conscious while others are not? Is it possible that an event can only be memorized if it is reported to and gains the attention of other neural modules? Associative memorization necessitates something that the item to be memorized can be associated with. This may be the previous percept within the same neural module, but if associative connections can be made to other modules as well then it is clear that the memory trace will be much richer and can be evoked more easily as the number of applicable cues is larger. The necessary intermodule connections can be established by unified inner attention so that the percepts about that specific event are broadcast to all other neural units. On the other hand globally available percepts constitute the contents of consciousness. Attention itself is controlled by significance and therefore it would be the significance of the situation that determines whether it would be globally broadcast and memorized or not, whether it would be available for retrospection.

Self-consciousness and Self-image

We are aware of our own existence. Self-consciousness involves the subjective sense of one's own physical and mental existence and the distinction between self and environment. In this way it also involves the perception of self as a vantage point, the perception of things to be out there and the bodily self that follows us no matter where we go. Self-consciousness involves also temporal continuity. We have personal histories, desires and expectations. We are the center point of our own life. We have a self-image consisting of these matters.

Fig. 8.1. Self-consciousness as self-image

As consciousness generally involves the perception of the physical environment, self-consciousness involves the perception of being physically distinct from the environment. Mental states are evoked externally by the environment as well as internally by earlier mental states or by bodily signals. Self-consciousness involves the ability to distinguish between externally evoked inner representations and internally evoked representations; to distinguish one's inner speech from somebody else's speech, to recognize internally evoked mental imagery as one's own fabrications as opposed to sensed imagery, and to recognize bodily needs as one's own.

Human physical self-perception is facilitated by the somatic system that consists of a variety of receptors in the skin and elsewhere in the body with fibers projecting to the somatic sensory cortex. The primary somatic pathway conveys information about touch, pressure and joint position. Separate pathways exist for pain and temperature[76,102]. It seems that the somatic fibers do not carry any information about the location of the receptor the only information being the strength of stimulation. Therefore the body image must be constructed by associating the various joint position signals, visual signals, touch signals, etc., together via experimentation. Touch and pressure sensors are important here. The distinction between the environment and one's own body can easily be detected by touching. When we touch external objects we get only one touch sensation but when we touch our own body we get two touch sensations, one from the touching part and another from the touched part. When these sensations are associated with the joint position signals and possible visual percepts the body image starts to emerge.

Mental self-distinction arises from the system's ability to make a distinction between the inner representations originating from sensors and those originating from inner processes. This distinction is necessary otherwise the system would react to its own inner imagery as if it were an actual external event. How can this differentiation be done? Internally evoked representations may be weaker, they do not contain all attributes of the respective external entities, e.g. inner speech is lacking direction and timbre. External signals have a location outside the system and this location is revealed by various cues, e.g. directional hearing and stereoscopic vision as discussed earlier. Changing spatial relationships between the body and the environment also help to make the distinction; whenever head is turned or position changed, the apparent location of the external signals change accordingly while internally evoked signals do not change. Another possibility to make this distinction is the observation of sensor activity. If sensors are not active while something is seen or heard then obviously these representations have inner causes only. The externally and internally evoked representations differ in a further way; internally evoked representations may be modified at will.

The perceived and recognized bodily self allows the formation of the concept "me", the entity to which personal history and intentions can be assigned. This "me" can be accessed via inner imagery where the imagined self is seen as the actor. This is especially relevant to planning as we must be able to "see" ourselves doing the planned action. The "me" can also be addressed verbally in inner speech. We can say, "Now I will do this". We can also say to ourselves "why did you act like this?"

The Theory of Mind or I Know What You Know and Feel

Self-consciousness allows a person to be aware of his own mind, its thoughts, emotions and knowledge. However, a person may also have the idea that others have similar minds and reasoning power; that they are able to think, deduce, remember and even deceive like us; that they feel pain and pleasure, have similar emotions and intentions as we do. This belief is usually called "the theory of mind". Unfortunately, in the literature "the theory of mind" is also used to refer to the general theory of cognition, while in this context we should speak about "the theory of the minds of the others".

The theory of mind responds to the questions: What is the other one doing, why and what is he feeling? What will the other one do when he has the knowledge that I know he has and not the superior knowledge that I have? The understanding of others' motives and desires will help us to interact with others and influence them. "Why is she behaving like this? Is she doing this to make me believe that..." The theory of mind involves also the realization that others have a theory of mind, too, and may use it to influence us or even to cheat us.

There may be a simple neurological function, mirroring by neurons, that can explain how a basic theory of mind can be acquired.

Imitation or mirroring is evident in human behavior. Small babies start to cry when they hear other babies cry even though they do not know why the others cry. People may start to laugh when they hear other people laugh even if they missed the joke. We may kind of feel pain when we see others hurt themselves. We hum to the tunes that we hear.

It has been found out that when we see or hear others doing something, the neurons that would be responsible for the execution of similar actions will be activated. These neurons are called "mirror neurons" even though mirroring is not their main purpose. The mirror neuron activation is largely subconscious and normally the actual response would be inhibited. However, the response may be acted out if the conditions allow that. The point is here, even subconscious mental mirroring of activity will evoke related associations, the meaning and purpose of the acts or merely a similar emotional state to that of the mirrored person. For instance, facial expressions are related to emotional states. When we see the facial expressions of others we may subconsciously mirror these and in this way become affected by the emotional states of others. Now there will be two possibilities. Either we take these emotional states as our own and act accordingly (obviously the primitive reaction) or we realize that these emotions are those of the imitated person. In that case sympathy may arise.

When a person is mirroring and imitating another person's actions and emotional states while the imitated person is observing this, positive feedback may occur. The imitation will amplify the actions and emotions imitated and this in turn will encourage the imitator. What we have here is the origin of the common spirit that can make teams excel. Unfortunately this kind of spirit is often demonstrated in inappropriate behavior. Youngsters may vandalize a location and afterwards can give no intelligent reason for their acts. Everybody will say: "I just did what the others did, I didn't start it".

Mirror neurons have already been discussed here in the context of learning by imitation. In part III of this book I will propose realizable mechanisms that learn and produce mirroring functions.

Mirroring does not explain higher theory of mind, however. The idea that others may use their knowledge to deceive us or that we may deceive others because we know what they know does not directly follow from mirror actions. Instead wider background knowledge and reasoning is needed. Deception can be learned by observing others that deceive. We can expect someone deceive us if we know that he has that kind of character and history and the conditions are suitable for deception.

The theory of mind allows us to understand the actions of others because we posses self-consciousness and realize that the others are similarly self-conscious and aware of their potential.

On the other hand a misplaced theory of mind may also be the source of anthropomorphism, the attribution of human-like mind or cognition to subjects that do not necessarily have these properties. Little children are easy victims to this: "My doll is so sad and hurt now".

Consciousness and Free Will

Free will and freedom of thought in human cognition manifests itself as the ability to make choices and to control one's thought flow. Free will also postulates the possibility of introspection and retrospection at will, therefore it is closely connected to consciousness. It seems that due to free will we can think any specific thought and have inner imagery any way we want to, we can make decisions as we please, we can choose our morals and values, we can choose our goals in life.

Also, it can be postulated that free will is not deterministic, because if it were then it would not be free.

It might be thought that what is not deterministic, must be unpredictable. Therefore if a machine had free will, its behavior would be unpredictable, unexpected. Is this really so? Many of us may know people, who by default have free will and especially one that happens to be against our free will. We know very well what they want, there is no unpredictability there. Perhaps these people have utilized non-deterministic processes in choosing what to want, but if we know a person we can fairly well predict how she makes her choices. So where is the unpredictability and non-determinism?

Let's consider free will in decision making. If a decision is made consciously then by definition the reasons for the decision are known and the decision can be mechanically derived from the reasons. This means that the decision has followed deterministically from the reasons and there has been no free will. Therefore it would seem that conscious free will cannot exist.

Benjamin Libet has reported about experimental findings that show that in the human brain the conscious intention appears only after a delay (350 ms) from the onset of a specific cerebral activity that precedes the voluntary act[53]. The original Libet experiment has been repeated many times since then. In a typical test set up the subject sees a kind of clock dial on a computer screen. The subject is then asked to push a button any time he pleases and to memorize the time point on the clock dial when the decision to push the button takes place. In the meantime the brain activity of the subject is monitored by an EEG measuring system. In each case it has been seen that an increasing neural activity has preceded the time point where the decision has been perceived to have taken place. It is as if the neural activity had to exceed a certain threshold before it could reach consciousness. Libet concluded that "voluntary actions may be initiated in the brain unconsciously, before any awareness of conscious intention to act, but conscious control of whether the motor act actually occurs remains possible".

What does this mean? The impulse to act, or to decide one way or another must come from somewhere. This impulse or cerebral activity has to spread out to larger scale activity that leads to the final action and it can only then become globally conscious when the spreading reaches other units.

How does this match with free will? It can be speculated that in a cognitive system parallel processes produce constantly a number of alternative impulses as a response to current conditions and internal needs. These impulses will compete for global attention and the strongest impulse will eventually win and initiate the related action or decision. It can be said that no actual free will exists here, the outcome is determined by the external conditions and internal needs, drives and emotional status even though the outcome may be impossible to predict in practice. However, this does not exclude the possibility of the perception of "free will". This "free will" may be perceived because the different options can be recognized in retrospect and the possibility of choosing differently is perceived. In this way the real "free will" may be illusory. However, because the choices depend on the status of the system's needs and drives in respect to the prevailing external conditions, it can be said at least that what is reflected here is the system's own will. Therefore it would seem that the impossibility of genuine "free will" would not exclude a system's "own will" that would guide the system's actions towards its goals and requirements.

Models for Consciousness

In recent years advances in cognitive neuroscience and computer technology have inspired various theories of consciousness. Such theories have been proposed for instance by Igor Aleksander, Bernard J. Baars, Antonio R. Damasio, Daniel C. Dennett, Gerald M. Edelman & Giulio Tononi, Edmund T. Rolls, Gerd Sommerhof, John G. Taylor and others. These theories are fruits of years of work and the brief summary that I can present here will not do the proper justice that they would really deserve.

Igor Aleksander argues that the brain is a state machine whose state variables are the outputs of the neurons. Therefore consciousness could be defined in the terms of state machine theory. Accordingly Aleksander proposes that consciousness arises in a suitable neural state machine via iconic neuron firing patterns. "The personal sensations which lead to the consciousness of an organism are due to the firing patterns of some neurons, such neurons being part of larger numbers which form the state variables of a neural state machine, the firing patterns having been learned through a transfer of activity between sensory input neurons and state neurons."[2] Aleksander has studied this by devising Neural State Machine Models "NSMMs"[4] and the Multi-Automata General Neural Unit System "MAGNUS"[1]. These neural systems are based on "weightless neurons" that act as binary input pattern recognizers. The operation of a properly programmed "weightless neuron" can be described with a truth table that indicates the input patterns that will cause the neuron to fire, i.e. to give "1" as the output. The neural system itself is a state machine.

Aleksander's "MAGNUS" system is a simulated neural unit system, which takes its input from another computer program "Kitchenworld". The input is in the form of simple figures. "MAGNUS" has two visual channels, one for objects and another that can be used to input names, etc., in image form. Obviously the system cannot directly handle words as strings of letters. The output of the system is used to control the position and size of the window over the picture delivered by the "Kitchenworld" program so that individual items can be imaged. MAGNUS is said to posses the property of state reconstruction; either the name input or the object input will lead to the reconstruction of an "iconic" state (binary pattern in the feedback loop) representing both. It is also suggested that MAGNUS is conscious because its neural firing patterns are meaningful in terms of its sensory world[15].

Bernard Baars' global workspace theory[12] proposes that the workings of the brain are widely distributed and the individual networks are controlled by their own aims and contexts. To organize the overall operation there is a network of neural patches that display conscious events, "a theater stage". Baars speculates that these areas are the sensory projection areas of cortex where the neural radiations from the senses first reach the brain. The "theater stage" is also a working memory, which contains the inner speech and inner imagery. The conscious experience created by the working memory is supposed to be under voluntary control. The conscious contents of these areas are broadcast globally throughout other, unconscious networks of the brain. Signals from outer and inner senses and long-term memory compete for access to the "theater stage" and this access is controlled by the "spotlight of attention". "Consciousness" is, according to this theory, supposed to broadcast a small amount of information to a vast unconscious audience in the brain. Baars describes consciousness as the "publicity organ in the society of mind". Baars' model includes also a director, "we — whoever we are", the observing self that has voluntary control over the theater stage.

The Baars theater and global workspace paradigm may be understood in terms of parallel distributed processing but it does not seem quite clear how the observing and controlling self is supposed to be implemented.

Antonio Damasio approaches the question of consciousness with a metaphor "movie-in-the-brain"[25,26]. This is a kind of multimedia show within our heads consisting of a large number of simultaneous tracks; sight, sound, taste, olfaction, touch, whatever senses there are. This constitutes our consciousness, or perhaps more accurately, the contents of our consciousness. Now Damasio sets to find out how this "movie-in-the-brain" is generated and how the sense of an owner and observer for this movie arises, especially how the owner and observer, "the self", appears *within* the "movie-in-the-brain".

Damasio proposes the following hypothesis: Brain structures map the organism and external world into second-order representation (the "movie"). This second-order representation occurs in thalamus and cingulate cortices. The sense of self emerges within the movie. Self-awareness is actually part of the movie and creates the "seen" and the "seer". Damasio sees that the biological basis for the sense of self arises from the parts of the brain that represent the continuity of the organism.

Daniel C. Dennett has proposed a so-called multiple drafts theory[29]. According to this theory consciousness is not based on a "theater" where it all would come together, instead consciousness is based on distributed processes. Dennett proposes that feature detections or discriminations are made once by specialized, localized portions of the brain without any master discriminator or observer. The varieties of perception and mental activity are under continuous editorial revision. These editorial processes are supposed to occur over large fractions of a second during which additions and overwritings of content can occur. In this way there is a constantly revised narrative stream of multiple drafts that may or may not be a part of the system's conscious experience. Dennett is not quite clear on the conditions of when and how consciousness arises from the process. Dennett's theory is not good at explaining the qualia or subjective phenomenal experience, which maybe is why Dennett goes towards denying the existence of qualia[30]. Dennett's multiple drafts theory is not really a neurological theory of consciousness, nor a computational one, instead it is more like a philosophical metaphor. However, the constructive point of this theory might be the emphasis on the importance of parallel distributed processing.

Gerald Edelman and Giulio Tononi propose that consciousness arises from certain material order arrangements in the brain[32]. Conscious experience is not seen to arise from the activity of any single area of the brain, instead it is associated with neural activity patterns occurring simultaneously in several neural groups in the brain, especially within the thalamocortical system. Additionally these neural groups must be engaged in strong and rapid reentrant interactions. Edelman and Tononi refer to the cluster of these neural groups as the "dynamic core" that is responsible for consciousness. Qualia are seen here as high-order discriminations among the states of the dynamic core.

What then is thinking according to the Edelman and Tononi theory? This theory does not really address that point and Edelman and

Tononi can only state: "What goes on in your head when you have a thought? ...We do not really know".

Edelman and Tononi subscribe to the idea of neural Darwinism; brain regions and their connections evolve competitively so that the connections that match the requirements best will survive and will be amplified.

Edmund T. Rolls proposes a linguistic or "thoughts about thoughts" theory of consciousness[85]. Rolls speculates that the brain language areas and parts of the prefrontal cortex are involved in control of behavior in a way that allows planning of action via language and syntactic manipulation of symbols. This mechanism would also allow higher order thoughts, thoughts about thoughts. Rolls sees that introspective consciousness would be the attentive, deliberately focused consciousness of one's mental states. Rolls notes: "Visual and memory mechanisms operate without consciousness. Consciousness involves higher-order thoughts. It is suggested that it would feel something to be a system that can think linguistically about its own thoughts and the phenomenal aspects of consciousness arise this way". This view clearly excludes consciousness from infants who cannot yet speak not to mention animals without language. If you cannot speak, you are not conscious!

Gerd Sommerhof argues that a theory of consciousness must cover the following three facets of the phenomenon: awareness of the surrounding world, awareness of the self as an entity and awareness of one's thoughts and feelings[94]. Sommerhof goes on to argue that consciousness can be explained by two key concepts and four propositions. The key concepts are the brain's Running World Model RWM and Integrated Global Representation or IGR. The Running World Model is a representation of the instantaneous external world, however including the body posture, movements and surface.

Sommerhof's four propositions are: (1) The brain forms an integrated representation of the current state of the organism and outer world. This is called the brain's Integrated Global Representation (IGR). (2) This Integrated Global Representation is subject to capacity and access limitations. (3) The Integrated Global Representation is the primary consciousness. (4) The subjective aspects of conscious experience or the qualia, consist of effects that an event has on the organism. These effects must be a part of the Integrated Global Representation.

In Sommerhof's theory the internal representations are constituted by the activity of sets of neurons.

Sommerhof states that self-awareness arises from the inclusion of representations of the current state of the organism into the Integrated Global Representation.

John G. Taylor proposes the so-called relational theory of consciousness[96,97,98]. Basically this theory maintains that consciousness arises from the interaction between perception and memory activities. Taylor states that "Consciousness arises solely from the process by which content is given to inputs based on past experience of a variety of forms. It has a relational structure in that only the most appropriate memories are activated and involved in further processing. It involves temporal duration so as to give time to allow the relational structure to fill out the input. This thereby gives the neural activity the full character of inner experience". Taylor elaborates further the relational point of his theory: "The conscious content of a mental experience is determined by the evocation and intermingling of suitable past memories evoked (sometimes unconsciously) by the input giving rise to the experience." Taylor gives an example: "The consciousness of the blue of the sky as seen now is determined by stored memories of one's experience of blue skies". The reader is asked to compare this to the perception process that I have outlined earlier.

Taylor uses "boxes-type" modeling of global information flow in the brain. Parts of the model are made more specific by proposing possible neural structures within the brain that could be able to perform the specified functions within the boxes. Taylor's two-stage and three-stage models propose general requirements for neural networks that are supposed to support phenomenal experience. These requirements include modular structure, localized representations in localized modules, good coupling between modules, temporal continuation of activity and suitable duration of individual activities. In the earlier two-stage model the first stage receives input stimuli and uses semantic level coding. The second stage utilizes buffered activity with lateral inhibition in order to single out the contextually appropriate representation. This representation enters awareness by being broadcast around the global workspace in the Baars' fashion.

In Taylor's three-stage model the first stages receive the input stimuli and are involved in low level feature analysis; the second stages are buffer stages that select the most appropriate percepts; the third stages possess attentional feedback to the second stages to refresh activity there and to the first stages to reduce or increase stimulus activity. There is feedback also from the second to the first stages. All stages contain a number of modules. The second stage modules are set to run a competition by lateral inhibition. The various feedback loops control the perception process so that obviously perception will be a function of external stimuli and internal states.

In Taylor's models consciousness (or perhaps more accurately the contents of consciousness?) is supposed to emerge as localized neural activity or "bubbles". This would seem to be something similar to Aleksander's "firing patterns" and Sommerhof's "activity of sets of neurons".

The Hammer Test for Consciousness Theories

How good is a theory of consciousness? What should a good theory explain? The problem here arises from the fact that different people see and define consciousness in different ways. For some consciousness may be more related to one's instantaneous subjective feeling of being here and now, others may see consciousness more like a some higher level thought process. Therefore theories of consciousness may also seek to explain different aspects of the phenomenon. But, are there some aspects that are more fundamental than others, aspects that every theory of consciousness should cover? How could we identify these aspects? Dr. Johnson refuted Bishop Berkeley's idealism by kicking a stone. Could we use a similar strategy to get on solid ground here? I think we can and I propose a method that I call the Hammer Test.

Take a big hammer and hit your thumb for all you're worth. (This is a thought experiment, the reader is advised not to do this in practice.) Now while seeing the stars and feeling the pain you can test whether your favorite theory of consciousness will survive.

The linguistic theory of consciousness is the first to fall. "Consciousness arises from verbal commentary of mental processes", right? You will realize that you do not need words to be conscious of the pain even though you might utter some words not really fit to be printed here. Or would it be that if you didn't swear, if you even didn't verbally think of it, you would not feel the pain?

Theories that propose consciousness as a social phenomenon will also fall. No interaction with others is needed to feel pain, but if there are spectators to your accident then maybe you might like to make a show out of it to get some sympathy.

Theories that specify memory as prerequisite for consciousness will also fall. "Signals become conscious when they are related to memories". However, you do not need much memory or evoked past experiences to be conscious of pain. You would feel the pain even if you were to forget its cause.

However, there may be higher aspects of consciousness and the content of consciousness that the hammer test does not address. Theories that fail the hammer test may still be useful when explanation for these higher aspects is sought.

Arguments against Machine Consciousness

Most arguments against machine consciousness are based on the belief that material machines cannot produce non-material mind. This is also the dualistic view — immaterial mind (the soul) and material body are separate substances. The basic dualistic argument is as follows: Consciousness (and cognition) is the property of immaterial and immortal soul (not the material brain), material machines do not have anything immaterial, therefore machines cannot be conscious. As dualism is not very fashionable among scientists today and the "immaterial soul" even less so, this background is often hidden.

There is a related "logical proof" from anti-materialism: Conscious machines are only possible if materialism is valid, materialism is a concept created by human mind and as such is not material; on the other hand materialism states that everything is material; as materialism is not material then materialism is wrong and conscious machines based on materialism are impossible.

Then there are some simpler arguments that all the same are based on anti-materialism:

- I am conscious, I am not a machine, therefore machines cannot be conscious. The logical flaw of this one is obvious, however this argument is often put forward in disguise!
- Only living things can be conscious, machines are not alive, therefore machines cannot be conscious.
- Transistors work like mechanical relays, mechanical relays are not conscious, therefore machines with transistors cannot be conscious.

Another line of arguments against conscious machines utilizes the supposed special status of man and these are from the Dark Ages:

- You must not undermine man's special position. Man is more than machine (or animal).
- Consciousness separates man from machines and animals, therefore machines cannot be conscious.
- You must not talk about conscious machines because it endangers our culture and morals. This is a good one! Do our morals depend on the evolution of technology? Would you teach your children that now it is okay to steal because we have, let's say, digital technology? I wouldn't and I don't.

PART III

TECHNOLOGY OF THE MIND

CHAPTER 9

MACHINE MODELS FOR COGNITION AND CONSCIOUSNESS

Introduction

Today it can be accepted that certain cognitive processes can be artificially reproduced. Pattern recognition, speech recognition, etc. are examples of electronically implemented processes that have found practical applications. These methods may not be perfect copies of the corresponding human processes of cognition, but nevertheless, in addition to their practical value they are also useful models for the study of cognition. It will be easier to understand aspects of human cognition if we know ways to produce similar functions artificially. It can even be said that we will understand cognition properly only when we are able to reproduce it at will. I would say that the same is very much true for consciousness, too. Machine models that could demonstrate even limited aspects of consciousness would be most valuable as they could provide tangible points of reference for discussion and study. This is a tall order, but nevertheless in what follows I will develop and propose machine models that could emulate or duplicate the processes of human cognition and even consciousness.

To what extent can this be done? Could there be any convincing arguments that a cognitive machine would possess any of the sought after properties? What if anything would the design of cognitive machines tell about the problem of consciousness? What should this machine be like and how would it differ from existing computers?

And finally, when all is said and done, would we only have some machinery able to do some tricks that are advertised as "intelligence", "thinking", "cognition", "emotion" or "consciousness" quite arbitrarily, without any real justification? I think we have seen examples of this and our deepest doubts are truly justified. So, how to address the problem properly and avoid these traps?

This is exactly why the previous discussion on the aspects of human cognition and consciousness is needed and I summarize briefly the main conclusions here. Thinking, as I have argued before, is not based on preprogrammed number crunching routines. The brain is not a computer that executes strings of command lines. Instead, as we have seen, human cognition is characterized by the flow of "inner speech", inner imagery, the basic cognitive processes like perception, attention, learning, deduction, planning, emotions, motivation, etc., and the awareness of these. Via these processes humans fuse and interpret instantaneous percepts from various senses according to their personal knowledge, goals and needs and in this way come to understand each instantaneous situation and its significance in the general context. This inherent understanding and awareness makes true cognition different from and superior to present day computerized intelligence.

Therefore, any true cognitive machine worth its while should also possess the processes of understanding and awareness. This capacity would make the machine philosophically interesting and, from a practical point of view, would also be most useful in the tasks of speech, scene, situation, text and story understanding.

As a first step towards the design of the cognitive machine the question of the representation and processing style of information must be addressed. The digital computer represents information by binary codes. These binary code words, consisting of strings of ones and zeros, are the symbols that the computer uses. All information processing within the computer is based on the manipulation of these symbols by syntactic rules. The symbols do not have any attached meanings for the computer, instead the result and response is solely determined by programmed rules. Human cognition utilizes also symbols, but please note, here is the difference. The symbols that humans use have attached meanings and these meanings are indeed used by the process. A symbol may even signify different things at different times depending on the general context. These things may also have varying significance and importance. It is exactly this processing with meaning that makes human cognition superior to any digital computer. Therefore a cognitive machine should also incorporate symbolic processing with meaning.

What would these symbols be like? There are levels of symbols. Obviously words and images are examples of symbols that we are aware of, but in the brain these symbols are carried by lower level symbols, the actual neural signals. These neural signals and processes are beyond our powers of introspection.

The important point is: These lowest level symbols themselves can be considered as carriers of the meanings only, their actual physical realization style would not be important as the physical nature of these symbols themselves would not contribute to the perceived meaning. It is the music, the modulation, that we listen to on the radio, not the carrier wave, be it FM or AM, even though without the carrier wave there would not be any music. Likewise the neural signals are only a carrier medium, one that allows the actual information to be carried and manipulated as kinds of surface features, the modulation. Thus it should not matter how the lowest level symbols are realized, be they signals in biological nerve cells or signals in an electronic circuit. This opens up the possibility for true cognitive machines. Electronics is what we master well and therefore an all-electronic implementation would be preferred. Thus we need to devise electric signal representations as the lowest level symbols that can carry the required meanings. Also we will need electronic circuit elements and assemblies that allow the manipulation of these signals in all required ways.

The requirement of attached meanings calls for means that can bind the lowest level symbols to external and internal entities, actions and relationships, that is, sensors and the processes of perception and learning.

Fig. 9.1 summarizes the complexities of a cognitive system, its own processes and connections to the environment.

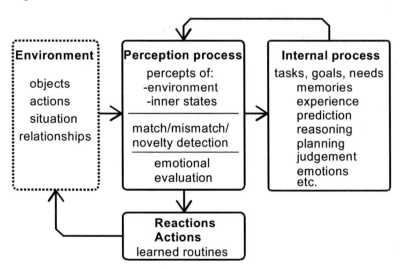

Fig. 9.1. The complexities of a cognitive system

The general purpose of the information processing in humans as well as cognitive machines is the production of meaningful responses; reactions, actions and communication. These responses have to reflect the demands of the tasks, goals and needs as well as the limitations and demands set by the instantaneous environmental situation. These demands are evaluated rationally and emotionally. Thus the elements to be considered here are the environment, perception process, internal cognitive processes, emotional significance evaluation as well as responses; reactions and actions including communication.

A complicated environment, conflicting lists of tasks, goals and needs as well as evoked memories and possibilities for inference would be a real challenge to any cognitive system. Purposeful operation calls for order and priorities, as everything cannot be attended to at once. Perception must be guided by instantaneous needs and importance. Likewise, the large number of internal cognitive processes that will have to run in parallel fashion must not be brought into attention as they are. Instead only the pertinent products of these processes should be brought into attention and used as intermediate results and starting points for further processing. This order can be achieved with the perception process that selects the external and internal entities to be forwarded to further processing as well as for the basis for reactions and actions. The internal entities — memories, intermediate steps and results of reasoning, etc. — must be brought into attention in such a way that their meaning can be interpreted by the system. This will usually mean that they must be represented in terms of perceived external entities, therefore these internal entities must take the form of inner imagery and inner speech.

Thus we can outline the general requirements for a cognitive machine. (1) A suitable method for the representation of information must be devised. (2) Suitable information processing elements that allow the manipulation of information by the chosen representation method must be designed. (3) A machine architecture that can accommodate sensors, effectors, the processes of perception, introspection and the grounding of meaning as well as the flow of inner speech and inner imagery must be designed. (4) The system design must also accommodate the functions of thinking and reasoning, emotions and language.

In the following chapters this kind of a model for a cognitive machine will be outlined. It utilizes distributed signal representation, associative neurons and neuron groups as the processing units and an architecture that supports multisensory perception, introspection, the flow of inner speech and inner imagery, language and emotion-like states and reactions. This model is based on the author's work on machine cognition[39]. The design will be described here in general terms that do not require specific background information. Some illustrations with boxes and arrays with clearly defined functions are presented.

The problem of consciousness will then be discussed within the framework of this machine model.

CHAPTER 10

REPRESENTATION OF INFORMATION

Representation of Information by Distributed Signals

The very first problem of machine cognition or any computing machinery for that matter is the problem of the representation of information. For digital computers numerical representation of information has been chosen. Therefore, for the computer all sensory information must be expressed by numeric values. Sound is represented by series of temporal samples, each having a distinct numeric value describing the instantaneous intensity of the sound. Images are represented by matrices of picture elements, pixels, each having a numeric value describing the intensity of illumination at that point of the image. The processing of the information thus involves algorithms that perform numerical calculations with these numeric values. These digital signal processing algorithms may nowadays be extremely complicated. But, as good as these methods are they have not taken us any nearer to real cognition.

Can we reproduce inner speech and inner imagery by numeric processing? I do not think so. Does the human brain utilize numeric calculations as the basis of cognition? Definitely not. So, if we do not accept numeric representation of information here then what else would be available?

We need a representation method that is flexible, one that allows easy description of the real world, its objects, entities and their relationships including action. This representation method should also allow easy modification and combination of these descriptions so that intelligence and imagination would be possible. This method should also allow imperfect representations, it should tolerate errors and distortion.

Associative processing seems to be a major component of human cognition. If we take this as a cue then the desired representation method should allow and support associative connections between representations and associative evocation of representations by other representations, even by incomplete ones.

Distributed representations have been proposed as an answer to these requirements[44]. Here a variation of this theme, one version of distributed signal representation is proposed. According to this method each entity is represented by a large number of signals that depict various properties of the entity. A visually sensed object may have signals such as size, shape, color, position, etc. A sound may be described by its constituent frequencies, loudness, envelope of amplitude, duration, etc. Each individual signal derives its meaning from the point of origin, from its own property detector. These signals do not represent their properties numerically, they only tell whether the designated property is present or not.

As distributed signals may be on or off, have the values one or zero, they may superficially remind binary coded patterns. However, it is worth noting the difference between binary coding and distributed signal representation. For instance the binary code 10110110 represents the decimal number 182 and this meaning is carried by the pattern itself because it is so defined by a universal convention. On the other hand detached distributed signal representation patterns do not carry meanings. The meanings are not defined by universal conventions. Instead, the meaning is bound to the physical signal lines and their physical points of origin. Therefore if a distributed signal pattern is transported to an arbitrary location within the same machine or another one, the meaning will be redefined by the new signal lines.

A simplified example of the principle of distributed signal representation is given in fig. 10.1.

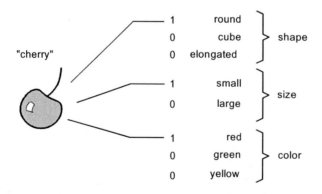

Fig. 10.1. Description of an object by distributed signals

In fig. 10.1 we have an object "cherry". We can describe this object by its shape, size, color and whatever we can detect about it. We can see that a cherry is round, small and red. If we have detectors for each of these properties then these detectors will generate the respective signals whenever the object is encountered. In this way this array of signals will indicate the presence of the object "cherry". This signal array is the distributed signal representation of the entity.

The meanings of the individual distributed signals should be chosen so that they are maximally independent of each other. If, for example, an object is described by shape and size signals and the size changes then the shape signal should not change if there is no actual change in shape.

We can see that the distributed signal representation allows easy modification of the appearance of depicted objects. Each property can be changed independently, we can make cherries smaller or larger. We can even assemble completely new objects, ones that we have never seen before by combining signals in a novel way. This property of distributed signal representation is useful when machine imagination and creativity is looked for.

Grandmother Signals?

How wide should a distributed signal array be? In the above example I had one signal to represent roundness, one signal to represent redness, greenness, etc. The example object "cherry" was represented by an array of these property signals. Obviously we could build a signal pattern detector that would output "one" whenever the signal pattern for "cherry" was detected. In this way we could represent "cherry" by one signal only. Could every possible entity be represented by one dedicated signal then? Indeed, this is a possibility and this representation method is sometimes called the "grandmother signal" method, as eventually there would be a single dedicated signal for the mental image of your grandmother, too.

If this method was applied to the representation of words then a dedicated signal for each and every word and every different form of the word would be needed. The number of the required signals would be a problem for any language and especially so for highly inflected languages like Finnish with its astronomical number of possible words.

Thus it is obvious that the use of several signals per entity is more practical. For instance the alphabet contains only a small number of letters, yet by these we can construct practically an infinite number of words. However, with signal arrays we cannot go down in the representational level as low as we wish, eventually there will be some basic entities that must be represented by one signal only. In the nervous system these entities are those that are detected by a single neuron. So, on the very lowest level single signal representations are necessary, but all the higher representations should be based on signal arrays of different widths.

In the following I every now and then pretend to represent also higher level entities by a single signal, this is for the sake of clarity only and in reality the use of multiple signals should be assumed.

Classification by Distributed Signal Representations

Entities can be classified according to their common features. In principle classification can be based on arbitrary criteria and there can be an infinite number of classes. In practice the criteria should be chosen by the requirements of the situation. Garments may be classified according to their color, cars may be classified according to their size, people may be classified by their age, profession, etc. Classification is not pattern recognition; it does not necessarily mean the recognition and grouping of closely similar objects, instead it means the identification of objects by some common property and therefore the detection of this property. In many cases objects must be classified on the fly, even objects that the classifier has not encountered before. Distributed signal representations are well suited for this kind of classification. The properties of the entities are already represented separately, therefore it is easy to see if a given entity has the requested property.

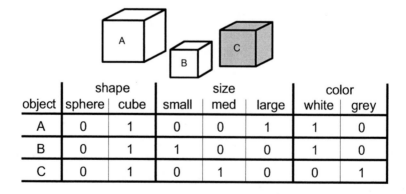

object	shape		size			color	
	sphere	cube	small	med	large	white	grey
A	0	1	0	0	1	1	0
B	0	1	1	0	0	1	0
C	0	1	0	1	0	0	1

Fig. 10.2. Classification; these objects are not identical, but they all are cubes

Fig. 10.2 gives distributed signal representations for three objects, A, B and C. From these representations it can be seen that each of these entities can evoke the class "cube" due to their "cube-like" shape features. Respectively the "cube" signal would select all these objects. In the same way "white cubes" would select the objects A and B, "small whites" would select the object B, etc. In this way distributed signal representations allow the classification of entities in multiple ways so that each entity may be a member of several different classes. This works both ways; when a class is given the members of this class can be evoked or when an entity is given, the relevant classes can be evoked.

Object Recognition and Alias Representations

According to the principle of distributed signal representation an object or entity is represented by a large array of individual signals, each indicating the presence of their specific feature. Thus, for instance a chair could be represented by signals that indicate the presence of four legs and some surfaces, fig. 10.3. Thus the array of these signals would symbolize the object "chair" and could be used whenever "chair" would have to be represented during information processing with distributed signals. Consequently, whenever the array of these signals is detected, the object "chair" would be recognized.

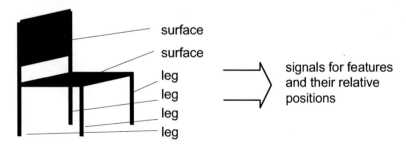

Fig. 10.3. The representation of a chair with distributed signals

However, things are not that easy. All chairs (or any other objects of a given class) are not similar. There may be chairs with three legs, one leg or with no legs at all. Thus we would get an ever-increasing array of signals whenever new examples of chairs were encountered. This is not really practical and therefore a more general description would be needed. For chairs this description could be: "A chair = a supporting structure & a surface to sit on". This is an interesting definition because the object, a chair is no longer defined primarily by its appearance, it is defined by application instead, the possibility to "sit on". Obviously the motor act "sit on" can be depicted also by a distributed signal representation and this representation can be evoked by the actual feature signals that describe the appearance of the chair as well as by any signals from the contextual situation.

Still, there are many ways to "sit on". There are many other kinds of applications as well and therefore the number of signals needed to represent a given entity may not always be economical on this approach. A more general representation with a consistent and minimal number of signals would be preferable. This kind of a representation that is used instead of actual feature signal arrays is called here the alias representation.

An alias representation is an arbitrary distributed signal array that is associated with an actual feature signal array which depicts a given entity. Alias representation can be considered as a label, token or a name that is used instead of the original signal arrays. The number of the individual signals in an alias representation may be much smaller than in the original signal arrays that the alias replaces. While the original signal array set may be poorly defined and fuzzy, the alias representation may be less so.

For instance the name "chair" may be used as the alias for the various examples and representations of the object "chair". Now everything that would be associated with "chairs" would be associated with the alias like all physical realizations, applications like "sitting on", etc. Eventually a network of connections will be formed and this network will reflect the same relationships as a network with the original feature signal representations, this time however, without the penalty of dragging along huge arrays of original feature signals.

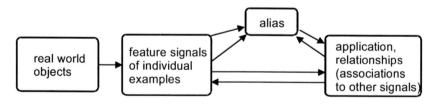

Fig. 10.4. Using alias representations instead of feature signals

How would alias representations be useful? When a distributed signal-based system perceives the external world large arrays of feature signals are always generated. If a considered response is to be generated then the system must consider different possibilities for action. This can be done by inner imagery, but now representations of remembered entities and relationships may be needed in addition to the perceived ones. This process might generate huge numbers of active signals if no signal reduction methods were used. Alias representation is such a method, it can be used to reduce the huge numbers of required signals.

One well-known alias representation system is of course the language. However, all alias representations need not to be linguistic ones.

The alias connection may be asymmetrical. A large number of property signals may evoke an alias representation, but the alias itself may evoke only a few of these properties. Recognition would be easier than reconstruction. This is the case also in the human brain; for instance it is easier to recognize a dollar bill than to describe it.

Representation of Significance

Associative processing with distributed signal representations can be a very efficient information processing method. However, in highly parallel associative systems a large number of distributed signal representations may be available at any given time. Therefore means must be provided to allow the pinpointing of most relevant representations so that these can be selected for further operations, to become the "focus of attention".

The meaning of a distributed signal representation is independent of the intensities of its component signals. Only the presence or absence of individual signals affects the carried meaning and a given signal array with high or low intensity signals depicts still the same thing. Thus the signal intensity is available for other purposes and can be used to carry significant information. Variable threshold circuits may then be devised for the selection of the strongest, most significant distributed signal representations.

In principle either continuous signals with variable amplitude or pulse train signals with constant amplitude can be used. If continuous signals are used then the signal amplitude (voltage) depicts the significance. If pulse train signals with constant pulse width and amplitude are used then the instantaneous pulse repetition rate can be used to carry the significant information because the repetition rate can be translated into respective voltage value by pulse integration.

CHAPTER 11

ARTIFICIAL ASSOCIATIVE NEURONS AND NEURON GROUPS

The Associative Neuron

Processing with distributed signal representations calls for the ability to associate representations with each other so that one representation can be evoked by another. Distributed signal representations consist of arrays of individual signals, therefore processing with distributed signal representations reduces to operations with individual signals. This basic signal processing unit is the artificial associative neuron.

A simple artificial associative neuron is a circuit element that is able to learn the association between an original signal and a number of other signals by repeated coincidences. After learning the associatively connected signals shall be able to evoke the original signal. This kind of an associative principle is also known as Hebbian learning[45].

This kind of a neuron can be realized quite simply. For each associatively connected signal we need a "synapse", a synaptic switch. This switch becomes permanently (or semi-permanently, depending on application) closed when learning takes place. Each neuron may have a very large number of synapses.

Correlative learning is assumed here; the original signal and the signal to be associatively connected must appear simultaneously several times (in special cases once only) within a period of time before learning can take place. A closed synaptic switch will then allow the evocation of the original signal by the other one. Thus the strengths of these switches have only two values, either the switch is on or off.

The strengths of the synaptic switches are not adjusted against each other, there are no special synaptic strength adjusting algorithms like the back propagation algorithm. This kind of a neuron is not numerical; it does not operate with numeric values and does not do calculations.

A signal derives its meaning from the point of origin, therefore the neuron must preserve the point-of-origin-path. The point-of-origin-path can be preserved if the neuron is configured so that the point-of-origin meaning is carried by one of the signals namely the original signal and this signal can pass through the neuron and become the output of the neuron with the carried over meaning.

Individual neurons can be assembled into neuron groups so that large arrays of signals could be associated to each other.

Neuron Groups and Memory Function

Distributed signal representations may appear as static or single parallel signal arrays that exist for a length of time. Distributed signal representations may also appear as temporal sequences, series of single representations, like the phonemes of a word or commands for motor acts like walking. Thus the following possibilities for cross-association exist:

single representation	→ single representation
single representation	→ sequence
sequence	→ single representation
sequence	→ sequence

Here a *neuron group* is defined as a unit that is able to learn all these four cases and will therefore be able to output both static representations and sequences as needed. It is obvious that this kind of operation cannot be achieved merely by taking a group of associative neurons, some short-term memories are needed in addition. However, possible technical realizations are skipped here, for our purposes this definition and the graphical depiction of fig. 11.1 will be sufficient.

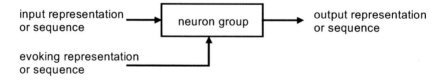

Fig. 11.1. The neuron group

The neuron group in fig. 11.1 is organized so that the evoking representation is associated through repeated coincidences with the input representation. Thereafter the evoking representation alone can evoke the corresponding output representation, which is the replica of the original input representation with which the evoked representation was associated. The operation is similar also for sequences. The evoking representation or sequence does not have to be exactly the same as the originally associated one. Distributed signal representation allows the use of "close enough" signal arrays as the evoking representations. This allows the generalization of learned associations; they can cover a large number of somewhat similar cases. If an exact match is not available then the closest match will be evoked automatically.

This definition and depiction of the neuron group is used throughout the rest of this book and whenever this depiction is used, the full possibilities of associations and evocations between single representations and sequences is to be assumed.

A special application of the neuron group is auto-association. Here the input representation is used also as the evoking representation. Thus, after learning, a part of the input representation can evoke the complete representation as the output. This works for sequences as well, a part of the sequence will evoke the remaining subsequent part of it.

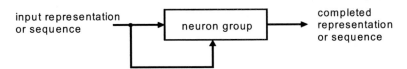

Fig. 11.2. Auto-associative connection

The realization of memory function is one important application of the neuron group. In principle the memory function consists of the storage and recall of a piece of information. In the digital computer memory is organized as addressed storage locations and the retrieval of the requested data, a binary word, is effected by activating the respective memory address. In an associative system however, no addressed memory locations exist and no memory addresses are used. Instead distributed signal representations are used and the recall operation is simply the activation of one distributed signal representation by another one. Thus the memory operation in a computer and in an associative system can be summarized as follows:

Computer: memory address → data
Associative system: representation A → representation B

Memory in the associative system thus involves the linking or association of distributed signal representations with each other. This is exactly what the neuron group does.

A neuron group stores memories in a kind of superimposed way. There are no individual memory locations for each memorized representation, instead the information of a number of memorized representations cohabit the same neurons in their synaptic strengths. Therefore the evocation of a desired representation by an associative input representation usually leads to at least a partial evocation of a number of undesired representations. These representations are usually evoked at lower signal level though and therefore an output threshold can be successfully used to suppress them.

Two useful output threshold strategies exist. (1) The common threshold level method: A common threshold level is applied to each neuron within the neuron group, all neuron output signals that surpass this level are allowed to pass. (2) The competitive winner-takes-all (WTA) method: The strongest neuron output signal within the neuron group sets the threshold and only the strongest signal is allowed to pass while others are inhibited. The strongest signal is usually also the best response to the instantaneous associative signals.

A neuron group may also be used to replace the associative signal array by another, alias signal array. In this case compression or generalization may be achieved if the number of signals in the replacing array is smaller than the number of signals in the associative array.

CHAPTER 12

MODELS FOR MACHINE PERCEPTION PROCESSES

Machine Perception

How do we make a machine perceive something? This is not merely a question of providing it with sensors like microphones and cameras. Sensors like these can provide raw sensory information, but the cognitive machine needs more than that. The machine needs to get to know what external entities are depicted by the raw sensory information and what is their importance. The machine needs also a way to perceive and introspect the results of its cognitive processes; the flow of inner speech, inner imagery and memories. The machine must also be able to utilize attention that is guided by expectation, context, needs, etc. A perception process model that provides these functions was outlined in chapter 4. Now we need to realize this model with associative neurons and distributed signal representations.

Sensory perception begins with sensors. Conventional sensors like microphones and video cameras may be used here, but the output signal that they provide is not suitable as such for these purposes. Therefore preprocessing that converts the sensor output information into a wide array of distributed representation signals is needed. Actual preprocesses depend on the sensor types, however for the purposes of this book no detailed descriptions of these are needed.

The actual perception process receives an array of preprocessed sensory signals and an array of feedback signals from the inner processes. Percept signals are determined by these and forwarded to the inner processes. Thus we will arrive at the perception principle[40] of the fig. 12.1.

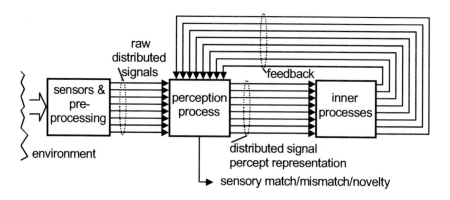

Fig. 12.1. The perception principle

The sensory and feedback information may be combined in various ways within the perception process block. Here one simple model is presented, fig. 12.2.

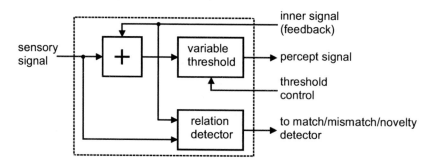

Fig. 12.2. Simple perception process model. This circuitry is needed for each individual sensory signal

In this model the intensity of each sensory signal is summed to that of the corresponding feedback signal. The resulting signal is then subjected to a threshold. If the intensity of the sum signal exceeds the threshold value, there will be output which now will be labeled as the percept signal. The threshold value is variable and controlled by the rest of the circuitry. If the threshold value is very low, either the sensory signal or the feedback signal alone may pass it and become the output, the logical OR function is realized. If the threshold is high enough then only the combined intensity of sensory and feedback signal may pass, both signals are needed and the logical AND function is realized. The threshold circuit does not affect the intensity of the passing signal. Therefore when sensory and feedback signals are summed, their combined intensity appears at the output; the sensory signal appears to be amplified. Thus the feedback signal can also be used to amplify (or prime) the sensory signal. The threshold function and signal amplification can be used as attentional mechanisms.

The perception process model of fig. 12.2 realizes the "doorbell" effect. The meaning of the percept signal is the same as the sensory signal, even when it is generated by the feedback signal only. Thus the feedback signal will be perceived as the external world property or feature that is represented by the corresponding sensory signal.

The perception process of fig. 12.2 determines also the relation between the sensory and feedback signals. This result is forwarded to another circuit that determines the match, mismatch and novelty conditions between complete distributed sensory and feedback signal arrays. Match signal is generated if the sensory and feedback signal arrays depict the same entity. Mismatch signal is generated if the sensory and feedback signal arrays do not match. Novelty signal is generated whenever a sensory signal array appears without a corresponding feedback signal array. Thus the perception process can, in principle, provide the information for questions like "is this a book (or any other known item)" or "have I seen this before".

In the following diagrams and later on in this book the numerous signal lines that depict the distributed signal connections in fig. 12.1 are replaced by a single line only for clarity.

Some special cases of perception can be distinguished. *Primed perception* takes place when the system is in some way predicting or searching for certain sensed representations. In this case the feedback signal array is a representation of the expected entity.

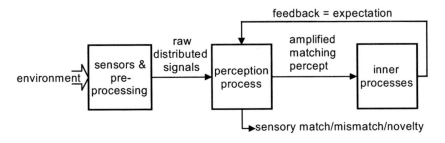

Fig. 12.3. Primed perception. Note: each line depicts a large number of signal connections

In fig. 12.3 a perception of an entity is expected. The entity given by the inner processes, the evoked representation, may be an object that is being searched for or it may be a certain word to be heard or even the tonal characteristics of the voice of a speaker that the system is listening to. Whenever the expected signals are sensed they will match the feedback and as a result they will be amplified by the perception process. Thus possible undesired signals will remain at lower level than the desired ones which can now be separated from the undesired ones by applying a suitable threshold.

There are two different cases of primed perception. *Predictive perception* takes place when the feedback representation is a prediction of the input. This prediction may arise due to associations to previous percepts. In predictive perception match/mismatch signals will indicate the accuracy of the prediction. *Searching perception* takes place when the feedback representation is an internally evoked representation of a desired entity to be found or distinguished. In that case the match/mismatch signals will indicate the successful/unsuccessful status of the search.

The perception process facilitates also the system's access to its inner states. This is *introspective perception* and in that case the system input, the perceived representation, consists of the feedback representation only, there is no sensory input or the sensory input is suppressed.

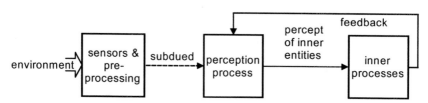

Fig. 12.4. Introspective perception

In fig. 12.4 the evoked output representation is imposed via the feedback on the input signal lines through the perception process. In this way the output from the inner process becomes a percept representation. Normally each individual signal on these lines derives its meaning from the point of origin, that is, the feature output of a sensor. The perception process preserves the point-of-origin-path and therefore the meaning of each sensory input signal is carried over to the percept signals. This original point-of-origin meaning is evoked whenever the signal lines are activated, also when the activation is caused by the feedback signals. The "doorbell effect" is thus utilized here: If the doorbell rings the impression is that there is somebody at the door even if the cause of the ringing was an electric short circuit somewhere along the wire.

For instance, in the visual domain each signal represents a visual feature. Therefore distributed signal representations activated by feedback represent also arrays of visual features, that is, imagery. From the system's point of view the situation is not really different from the situation where actual external objects are visually sensed. However, in this case this imagery has its origins within the system itself, it is inner imagery. These images are not necessarily those sensed before, instead novel feature combinations are possible due to the nature of the distributed signal representation. In this way the perception process allows the perception of system's own inner products and the grounding of the meaning of these in the external world entities.

The introspection of inner linguistic thoughts is facilitated in the same way in the auditory domain. The feedback representations are now imposed on signal lines that carry auditory features. Therefore the feedback representations are perceived as something that is heard; the introspected inner linguistic thoughts are transformed into "heard" speech. Thus the machine would perceive its linguistic thoughts as "inner speech" just like we humans do.

However, not everything can be introspected. The actual processes and intermediate results will remain beyond introspection because the feedback loop returns only the instantaneous final results to the perception process.

This basic grounding of meaning for sensed and introspected percept representations is a key point and essential to the understanding of the cognitive architecture that will be introduced in the following chapters.

Sensory Modality Perception/Response Modules

An actual cognitive system would include a number of sensory modalities like visual, auditory, tactile, etc. The perception process principles should be applied to each of these modalities separately. Thus each sensory modality would have its own preprocesses, own neuron groups and own perception process blocks; a complete perception process module of its own. As these modules are also able to generate responses that they can introspect via the feedback loop, I will call them perception/response modules.

These sensory modality perception/response modules constitute basic system modules that within each sensory modality realize the functions of direct perception, primed perception, introspective perception and the establishment and the grounding of meaning of inner representations. The graphical depiction of fig. 12.5 for the perception/response module will be used throughout the rest of this book.

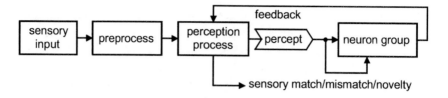

Fig. 12.5. The sensory modality perception/response module

In fig. 12.5 the sensory input would consist of suitable sensors like microphones for auditory modality, light or image sensors for visual modality, touch sensors for tactile modality, etc. The preprocess would transform information from these sensors into suitable distributed signal representations so that a large number of signal lines would enter the perception process block. Each signal line would carry only the information of the presence or absence of its own feature; this would be the basic meaning for the signal. The perception process would not alter these basic meanings. Only the signal strengths could be changed and weaker signals could be rejected.

The output of the perception block would depend on the actual sensed information and feedback signals as described before. The signal array at the output of the perception process block is labeled as the percept because it is the instantaneous final product of the perception process. The percept is the "official" output of the module and is transmitted as such to other modules.

You will notice that in fig. 12.5 the percept is connected to the neuron group in two ways, directly and as an evoking signal. This constitutes the auto-associative connection of fig. 11.2 and thus the perception/response module will be able to complete partial percepts of objects that are, for instance, obstructed by other objects or are seen in bad illumination, are heard poorly because of noise, etc.

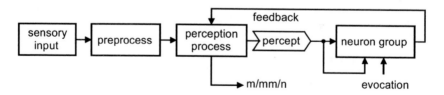

Fig. 12.6. Evocation input at the neuron group

The neuron group is able, as defined before, to evoke representations as a response to evoking signals. Thus an evocation input can be included in the neuron group. Via this input a "need" representation would be able to evoke a respective sought-after representation, which would be fed back to the perception process and matched there against the actual sensory input.

Match, mismatch and novelty signals are generated by the perception process. The match signal indicates that the sensory signal array and the feedback signal array match, the internally evoked item and the sensed item are the same or almost the same. The mismatch signal indicates that the internally evoked item is not sensed. The module novelty signal indicates that no internal evocation is available, therefore the sensed item is a new one.

The feedback loop can be made to circulate its signal for a while so that the instantaneous percept is sustained temporarily. Thus the perception/response modules can also act as reverberating short-term memories.

Cross-Connected Perception/Response Modules

A perception/response module by itself is only able to handle representations of its own sensory modality, it can only associate its own types of representations with themselves. Without any reference to other modalities these representations can only have basic meanings that are determined by the point-of-origin feature detectors.

Higher cognition arises from the possibility that representations of entities can be made to symbolize multiple affairs via multiple connections to other representations. The power of language does not come from the fact that a sensed signal array caused by an uttered word signifies the respective basic point-of-origin meaning, the sound pattern. The power of visual cognition does not come from the fact that a certain sensed signal array signifies a certain visual pattern. Instead, the power of language comes from the fact that a signal array, which corresponds to a sensed sound pattern, can be made to represent something completely different. Via further associations to other representations this signal array can turn into a word that signifies the associated entity. Thus instead of mere sound patterns we get meaningful words, instead of mere visual patterns proper objects with relationships to other entities can be perceived.

Therefore, in order to facilitate the inclusion of these associative meanings, associative cross-connections between different perception/response modules are needed.

These cross-connections can be realized via the in-loop neuron groups as depicted in fig. 12.7.

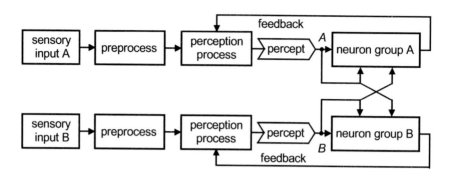

Fig. 12.7. Cross-connected perception/response modules

In fig. 12.7 the perception/response modules transmit their percepts to each other's associative inputs at the neuron groups. These cross-connected modules will now be able to learn sensory input entity pairs. If a certain percept *A* and another percept *B* appear together repeatedly, they will be associated together. Thereafter the percept *A* will be able to evoke the constituent signals of the percept *B* at the neuron group of module B. These signals are forwarded to the perception process of the module B and will thus become the percept *B*. Likewise the percept *B* will be able to evoke the percept *A* at the module A. The associative evocations amplify each other and a self-sustaining closed loop is established; the percepts *A* and *B* are bound together.

It should be noted that the percepts *A* or *B* alone or some new percepts that are rather similar to *A* or *B* are able to give rise to this kind of binding. In this way a module is able to perceive or "interpret" in its own terms what ever the other module is perceiving, what ever is the instantaneous result of the other module's activity.

During the cross-associative evocation there might already be other output representations pending, for instance reverberating percepts. In that case the evoked and already existing representations would have to compete against each other at the neuron group output threshold. The winner would be passed and directed to the perception process.

Match/mismatch detection can be included in the neuron groups. The m/mm operation at the perception process judges feedback signals against sensory input only. However, deduction by internally evoked representations only, without any incident sensory information calls also for match/mismatch detection. This can be achieved by the neuron group match/mismatch process that judges percepts against percepts even if these were internally evoked.

CHAPTER 13

A COGNITIVE MACHINE ARCHITECTURE

Integration of Processes and Functions

Our awareness of the environment does not consist of separate visual, auditory, touch, etc., sensations. Our cognition is not a collection of separate mental concepts either. Instead, at each moment our percepts from the various sensory modalities are bound together. A sound is connected to a visually detected source, a sensed object is connected to mentally perceived possibilities of use. The process of learning to recognize visual objects does not rely only to visual percepts but also percepts from other sensory modalities, for instance touch sensations. Later on these cross-connections between the various sensory modalities provide us with the possibility to instantly interpret percepts from one sensory modality in terms of other ones. Not only are these connections made between actual sensory percepts, also needs and memories are coupled in. In short, our instantaneous percepts, needs and ideas are bound into a stream of mental contents, the contents of our consciousness; inner imagery, inner speech and sensations. We can also act upon the environment; we can move around, we can pick up objects and do various things with them. This necessitates the integration of sensory information, the inner imagery of planned action and the actual motor commands that initiate and control motor action.

Likewise a cognitive machine should be able to combine sensory information from various sensors and information originating from memory and the instantaneous needs of the machine. A true cognitive machine should also have the flow of inner imagery and inner speech with properly grounded meanings. The machine should also be able to exercise attention, to select the relevant information and train of concepts at each moment.

Thus a machine architecture where stand-alone sensory modules are connected to some central processing unit is not satisfactory. Instead a machine architecture that allows the above kind of integration of information and operation must be devised.

Here the principle of the cross-connected sensory perception/response module pairs is a good starting point. However, this principle needs to be expanded here as a complete cognitive system would have to contain not only two but many sensory modalities. The general principle will now be: Each sensory modality has its own perception/response module. These modules shall be cross-connected so that percepts from every module can be broadcast to every other module. Local thresholds at each module are needed to determine whether the broadcast percept will be accepted by the receiving module or not. In this way we will get the architecture of fig. 13.1.

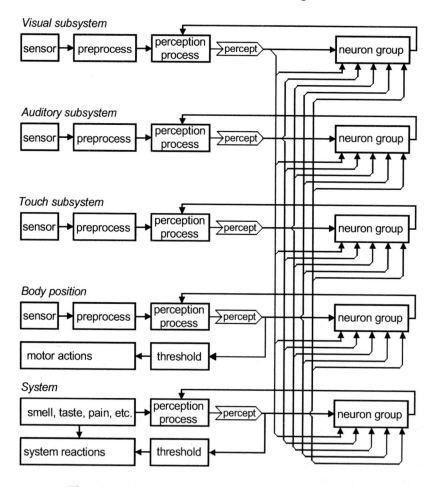

Fig. 13.1. The overall cognitive machine architecture

In the architecture of fig. 13.1 the sensory modalities of vision, audition and touch are depicted. A robotic application calls also for a body. Therefore sensors for the instantaneous positions of the various body parts are also included. The machine executes physical work through the motion of these body parts. The planning of any motor action involves inner body position imagery of the required body part position sequences. The relevant motor commands must be controlled by these so that the intended action can be executed. Therefore the motor command units are connected to the body position sensory modality.

The machine needs also sensory modalities that are able to produce sensations with intrinsic good/bad value and significance. These sensations would be the machine equivalents for smell, taste, pain and pleasure and these sensations would be able to initiate instant system reactions like rejection, acceptance, approach, avoidance, etc. These sensors and system reactions would constitute the basis for machine emotions as will be discussed later on.

Basically each of these modules in fig. 13.1 are doing their own thing, making percepts of their own kind from sensory information and their own predictions of these and broadcasting the instantaneous percepts to other modules. However, they also receive information from other modules and this information may affect the operation of the receiving module by evoking new representations there or by amplifying some representations that are already active. In order to have any effect the broadcast information must pass thresholds at the neuron group associative inputs. These thresholds pass the information from the module that broadcasts with the strongest signal intensity. These thresholds may also be set so high that no information is accepted at that time.

Sensory and inner attention are realized here by the control of signal intensities and threshold levels. This allows the realization of the principles of attention that were discussed in the chapter "Attention in Perception and Thinking".

It can be seen that this architecture utilizes self-repeating universal neural structures, which are similar to each other for each sensory modality. Moreover, more detailed inspection would reveal that these structures need to repeat themselves within every modality. This leads to design economy, as only one kind of neural building block needs to be designed. Because of this there is also a possibility for dynamic circuit resource allocation. These self-repeating neural circuits need not to be earmarked initially for any sensory modality. Instead the system could be organized so that during learning each sensory modality recruits as much as necessary of the common circuit resource, of course within the limitations of the total circuit capacity.

However, this general machine architecture idea is not as yet sufficient for the discussion of machine cognition and consciousness. We will have to examine in more detail the workings of the various subsystem modules, especially vision and audition. Also motor acts and system reactions have to be considered in more detail. It is useful to consider here an application where the cognitive system controls a non-specific robotic machine so that references to motor responses and the hardware body can be made.

Visual Subsystem

The visual subsystem has three main tasks: (1) Visual detection of objects in the environment, estimation of their appearance, shapes, sizes, position and motion. (2) The generation of the visual vantage point reference; the determination of the machine's position in relation to the environment. (3) The generation of inner imagery; visual imagination. The visual subsystem is also cross-connected to other sensory and motor modules allowing the integration of visual information to the other sensory information and to the planning and execution of motor tasks.

An image sensor, somewhat similar to the human eye, is assumed here. This sensor would consist of image forming optics, a lens, and an image sensor consisting of a matrix of sensor pixels. The lens would project the image on the pixel matrix, which in turn would output the illumination value of each pixel in the form of electric voltage signals. The pixel density would determine the resolution of the image sensor. Here it is assumed that a small center part of the sensor pixel matrix is made denser and will thus have higher resolution. This center area would correspond to the fovea of the human eye and accordingly will be referred to later on as the (artificial) fovea. Also, what has been earlier said about the effects and benefits of the limited size of the fovea will apply also here. The gaze direction and the fovea are related; when the gaze is directed towards an object, the image of the object will be projected on the high-resolution fovea. Existing solid state image sensors can be used here successfully.

The actual visual detection of objects involves the separation of the objects from the background and from each other; binding the various visual attributes of an object together and preserving the integrity of the object during illumination change, temporary obstruction and motion of the observer and the object itself. Also directions and distances must be estimated. To satisfy these requirements distributed signal representations are used along the lines outlined earlier.

The output from an image sensor is a map of pixel illumination intensities and as such does not constitute a proper distributed signal representation. Therefore preprocessing is needed to derive true distributed signal representations from these, like those for shape, size, color as well as motion and position information. These representations shall be independent of each other. For instance, shape information shall not contain size, color or position information that is, any given shape shall be represented by a respective representation which is not affected by the size, color or orientation of the object. It should be especially noted that the distributed signal representations for each class of percepts; shape, size, color, etc., are superficially similar to each other; they are just arrays of signals. The meanings of these signals are not carried by any special signal property. No signal here is "round", "large" or "red" in itself, these are not the properties of these signals. Instead, each signal represents the property detected by the point-of-origin feature detector. If, for instance, the feature detector at the front end of a given signal line detects wavelengths that correspond to the color red, then the signal on that specific line indicates the presence of the attribute "red". Thus the detected features or properties are carried as the modulations on the signals, either on/off or graded.

Detailed descriptions of these visual preprocesses are not needed here. In what follows it suffices to assume that suitable distributed signal representations of the visual information can be produced by these preprocesses. Thus individual perception/response modules can be assumed for shape, size, color, motion and position information and we arrive at the architecture of fig. 13.2.

Fig. 13.2 is a more detailed description of the visual subsystem, which was shown only as a single loop in the overall architecture in fig. 13.1. This architecture utilizes also the principles of cross-connected perception/response modules as discussed before. The cross-connection between the individual loops allows the binding of shape, size, color, etc., information so that the various signal patterns that describe these properties for the sensed object can be manipulated as a whole.

The visual subsystem shall also generate the visual vantage point reference; the system shall "perceive" itself as the center-point surrounded by the environment. The objects sensed by the visual sensors must not remain as excitations on the sensor pixel matrix, instead they must be perceived to exist out there, in the external world. For this purpose the visual subsystem shall have means to determine the direction and distance of the observed objects, their position relative to each other and to the machine. Moreover, there must be connections to the motor system. The system must be able to determine what it would take to reach out for a visually perceived object, what it would take to travel towards it in terms of motor commands.

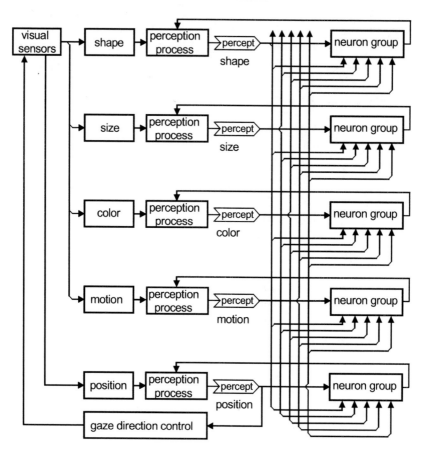

Fig. 13.2. The visual subsystem of fig. 13.1 in more detail

The limited size of the artificial fovea can be used to determine the direction towards an object. When the machine "gaze" is directed towards an object, the object image falls on the fovea and will be perceived with maximum resolution. Thus controlled gaze direction will be instrumental in the determination of directions. Here a predefined line of reference will be useful.

In order to illustrate the aspects of the visual vantage point reference a robotic machine is assumed here with a moving body and a movable head, as shown in fig. 13.3.

In fig. 13.3 two visual sensors are attached to the head in a way that allows a small movement relative to the head. In this way "gaze direction" can be controlled by moving these sensors. Now a line of reference can be defined. This line points straight forward from the head and allows thus the definition of relative positions like "straight ahead", "up", "down", "left", "right", etc. The gaze direction determines the direction of a visually attended object and also its position in relation to the line of reference. The line of reference does not change when the gaze direction is changed while the head remains stationary. However, when the head turns, the line of reference changes respectively because the direction "straight ahead" changes. The turning of the head will allow the machine to map directions towards objects all over.

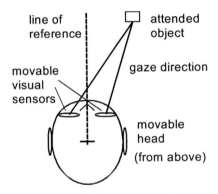

Fig. 13.3. The line of reference determines the "straightforward" direction in relation to the robot head

A separate sensory loop is provided for the perception of the relative position of visual percepts as shown in fig. 13.2. This position is determined by the respective gaze direction and head position. Thus the percept of the instantaneous direction is associated to each instantaneous set of shape, size and color percepts and vice versa. Thereafter the evocation of any given direction will evoke the respective set of shape, size and color percepts; the object that was perceived at that direction. Likewise, any given object will evoke the direction of that object. Thus the machine will "know" also the relative positions of those objects that are currently outside the visual field. This kind of association between visually perceived objects and their positions will constitute a kind of "inner map" of the current surroundings. This association of objects and their respective positions must of course be temporary. Any new percepts must be able to overwrite old associations, the revision of the "inner map" must be possible whenever the surroundings change. Thus the visual vantage point reference has been established. The objects are out there with defined relative positions and the machine "knows" in terms of motor commands how to reach them.

The "inner map" mechanism may also be used as *a temporary store for inner imagery*. Imagined entities may be positioned into virtual positions by assigning them a virtual gaze direction. Thereafter imagined entities may be brought into attention by directing the virtual gaze direction as needed. It should be noted here that full resolution is not needed for the "inner map" or "inner imagery store". In order to minimize the required storage capacity, that is, the number of neurons and synapses, only the "outline" or a rather sketchy description of the visual objects need to be stored. This is obviously how we humans do it, too.

The visual subsystem shall also be the seat and vehicle for the flow of inner imagery and visual imagination. The essential problems of inner imagery and visual imagery relate to the evocation of the required visual feature signals, the introspection of these and the grounding of meaning for these.

The principles of introspective perception are used here. Each signal line originating from the visual sensors and feature detectors carry a signal that represents a visual feature. During introspection a set of these signals are activated by the feedback from the visual neuron group. The evoked signals depict visual features due to their hardwired origination path and therefore these signal patterns correspond to imagery, which in this case has internal origin. This imagery does not have to possess the photograph-like quality of actual visual percepts. We humans do not usually have this, either. Instead visual imagination can take place as the manipulation and combination of merest essential visual features in the form of feature signal arrays. These feature signal combinations would correspond to very sketchy visual ideas, but these would still be sufficient to carry the required meanings and to evoke the necessary associations at other sensory modalities.

The images of inner imagery are not necessarily those sensed before, instead novel feature combinations are possible due to the nature of the distributed signal representation. Unicorns and other novel images can be created by taking already known visual features, making them larger or smaller, combining them in new ways, etc. This is all facilitated by the distributed signal representation that allows the adjustment of the visual features of an object independent of each other.

The cross-connections to other sensory modules will enable the evocation of inner imagery by other module percepts and vice versa. Examples of these will be given later on.

Auditory Subsystem

If we close our eyes we will still be aware of our surroundings. We will still perceive ourselves as the center point surrounded by the environment. This time, however, the environment manifests itself as an auditory landscape consisting of individual sounds coming from various sources and directions. Thanks to audition we are able to locate the sources of the sounds and are able to approach or avoid these. Due to cross-associations to visual and tactile percepts we are able to recognize objects and phenomena like cars, airplanes, cats and dogs, doors, footsteps, etc., by their sound signature alone.

A cognitive robot should be able to perceive the auditory environment in the same way. The auditory subsystem should be able to resolve individual sounds and the locations of their sources and be able to focus attention to any of these sounds. Individual sounds are not temporal snapshots. Instead they have also a distinct duration as well as time-varying qualities; they are temporal sound patterns. The auditory subsystem should also be able to handle temporal sound patterns as entities that can be associated with and evoked by each other and other sensory modality entities. This is also a prerequisite for a spoken language.

Microphones can be used as the auditory sensors. However, the output from a microphone is a time-varying voltage that follows the intensity variations of the sound pressure. This time-varying voltage contains the sum of all instantaneous sounds within the microphone's reach and therefore is not suitable as such for cognitive processing. Therefore preprocessing is needed so that individual sounds and sound patterns could be resolved. For that purpose instantaneous frequency components and their intensities can be derived from the microphone signal. Also rhythm patterns can be detected. This information can be represented by distributed signal arrays. Once more it must be emphasized that the distributed signals do not have the frequency or duration that they represent. Instead they only indicate the presence of a given frequency or the duration of an instantaneous sound pattern or interval.

In order to segregate individual sounds we must know which frequency components belong to the same sound and should be grouped together. The principles that were presented in chapter 4 can be used here. Frequency components that share a harmonic relationship usually belong to the same sound and may thus be grouped together. Also frequency components that appear at the same instant most probably belong to the same sound and may thus be grouped together. The direction of individual sounds can be resolved by using two microphones. Thus the auditory perception process would consist of the perception of sound features such as frequency components and their temporal behavior, rhythm perception and direction perception. Separate perception/response modules can be assigned for these and we get the architecture of fig. 13.4 for the auditory subsystem.

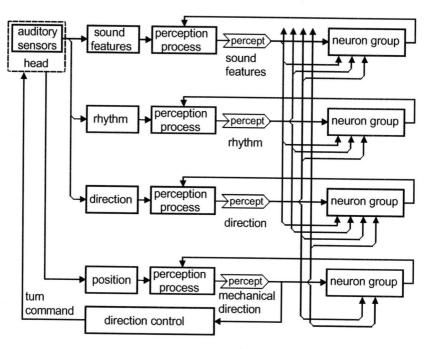

Fig. 13.4. The auditory subsystem of fig. 13.1 in more detail

In fig. 13.4 the separate modules for sound features, rhythm and sound direction are cross-connected so that the properties of a given sound can be bound together.

The auditory vantage point of view requirement calls for processes which enable the system to "perceive" the origin of the sensed sounds to be out there instead of the actual point-of origin, that is, the microphones themselves. This can be done because there is nothing that would fix the origin of the sound to the vibrating membrane of the microphone. The required processes are the detection and estimation of the direction and distance of the sensed sounds and their association with visually sensed objects.

There is one important cue that gives away the external point-of-origin of a sound. When we turn our head, the sound direction changes in respect to the head. If the sound had internal origin then the turning of the head would not cause any changing direction sensation.

In order to utilize this principle in our robotic application it is supposed that the auditory sensors, microphones, are fixed to the robot head which can be turned. The head direction must now be sensed and this is accomplished by the head position sensing module as indicated in fig. 13.4. On the other hand the sound direction is detected by the use of two microphones. Let's suppose that the sound source is straight ahead. Now the straight-ahead line of reference of the head and the sound direction as sensed by the microphones will coincide. If the head is turned, the sensed sound direction will change in respect to the line-of-reference. However, this deviation will match the sensed head turning. Thus it can be inferred that the environment has been static and the apparent change in sound direction has been due to the turning of the head only. The position of the sound has not followed the head when it was turned therefore the head and the sound source must be separate.

The auditory sensing of sound source location is not always very accurate. However, the perception of sound source locations can also be assisted by vision. Therefore the robot system must also be able to associate visually detected objects and any sounds that they give out. For this purpose the sound direction perception shall be cross-connected to the head position module so that the head could be turned towards any sound. This would allow the focussing of visual attention to the sound source. The simultaneous focussing of attention to the auditory and visual qualities of an object would allow the association of these. Afterwards the objects and their sounds would be associatively connected. If a typical sound of an object were heard and the corresponding object were seen then the point-of-origin of that sound would be taken to be the same as indicated by the visual system.

The association of auditory sound patterns and visual entities has further implications. A perceived sound of an object will become a token that stands for the object; it will be able to evoke an inner image of that object. The sound of a percolator next room may evoke the image of that machine, furthermore it may evoke the possibility of having a cup of coffee, etc.

Now we must ask: Associative systems are tricky and almost everything may be associated with everything else; can we design the system so that only those sounds that are generated by an entity are associated with it? Obviously it would be very difficult to ensure that a sound which appears simultaneously with an object is indeed generated by that object. Thus it may well be that the association of sounds and visual entities remains imperfect; sounds that do not actually belong to an entity are associated with it anyway.

What are the consequences of this shortcoming? Should we raise our hands and admit failure? No. Instead, it is exactly this imperfection that allows a tool of higher cognition; language. Due to this imperfection sound patterns, words, that actually have no natural connection to visually perceived objects and acts or percepts from other sensory modalities may nevertheless be associated to these and vice versa. The auditory subsystem, once designed to be able to handle temporal sound patterns will also be able to handle words, their sequences and their associations. Thus no further requirements for the acquisition of words are needed, this ability to use certain sound patterns as symbols for other entities arises not from elaborate additional circuitry but from the imperfection that is inherent in the associative process.

Indeed, one important function of the auditory subsystem is the facilitation of language and inner speech. These functions will be discussed in more detail later on.

Visually Assisted Auditory Perception

The auditory and visual sensory modalities are cross-connected as depicted in the overall cognitive machine architecture picture, fig. 13.1. This allows the association of auditory and visual entities and the evocation of one by the other. The cross-connection allows also visually assisted auditory perception. Here visual percepts may create expectations for auditory perception leading to enhanced perception under noisy conditions. The relevant cross-connections during visually assisted auditory perception within the cognitive architecture are depicted in the following simplified picture that outlines the cross-connected auditory and visual perception/response modules.

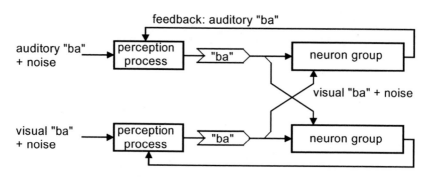

Fig. 13.5. Signal flow during visually assisted auditory perception

Visually assisted perception of speech is taken as an example. Here the system perceives visually the speaker's face and lips and receives the spoken words and syllables via the auditory sensors. Let's suppose that this process takes place under noisy conditions. The actual spoken syllables, say, "ba", will be masked by noise and the visual perception may be degraded by poor illumination. The visual modality perceives now lip and mouth motions that more or less correspond to the syllable "ba". These visual percepts are transmitted to the auditory neuron group where they evoke the respective auditory representation of "ba" plus noise. Some of the noise contained in the visual distributed signal representation is rejected at the auditory neuron group by the associative evocation process that tries to evoke the best match for the input signal. The evoked auditory representation is then forwarded to the auditory perception process where it is summed to the sensed auditory representation. This situation is depicted in the fig. 13.6 where on the left-hand side the intensities of individual input signals are given. Solid lines depict sensed auditory signals and dotted lines depict feedback signals.

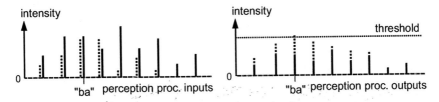

Fig.13.6. Perception process input signals and output signals. Left: solid lines = sensed signals, dotted lines = feedback signals. Right: the output is the sum of sensed signals and feedback signals

The perception process sums and scales the respective sensed signals and feedback signals giving the output signals of the right hand side in fig. 13.6. It can be seen that the correct perception of the syllable "ba" will be now achieved while the sensed auditory or visual signals alone would not suffice. This improvement is due to the fact that the sum of non-correlated visual and auditory noise signals tends to be less than the sum of coherent true signals.

The McGurk effect[57] (see chapter 4, section on "Multisensory integration") may arise when the visual and auditory percepts do dot depict the same auditory entity. In that case the perceived sound may be different from the actual heard sound or the visually perceived sound. McGurk-type effects may also result here if the auditory and visual inputs do not represent the same thing. In a typical example of the McGurk effect visually perceived pronunciation of the syllable "ga" and simultaneously received auditory "ba" lead to the auditory perception of the syllable "da".

Lip and mouth activity that corresponds to the syllable "ga" is perceived by the visual sensory modality. This percept is transmitted to the auditory neuron group where the respective auditory representation for "ga" is evoked. This evoked representation is forwarded to the auditory perception process via the feedback loop. Now the actual auditory percept will be the one that arises from the combined neural representations of the auditory sensation "ba" and feedback "ga". The visual percept cannot now amplify the auditory percept, so what would the resulting percept be?

As stated earlier, the perception process simply adds the sensory representation and feedback representation together. Again, all of these representations are groups of neural signals centered around the strongest signal that best describes the "ba" and "ga". If these groups overlap, that is, they share some signals, then these signal intensities will be summed and the new strongest signal would not be "ba" or "ga", instead the strongest signal would be found in between these signals, fig. 13.7.

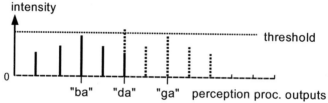

Fig. 13.7. The emergence of the McGurk effect at the auditory perception process

In fig. 13.7 solid lines depict sensed auditory "ba" signals and dotted lines depict visually evoked "ga" feedback signals. Individual neuron signal intensities are the sums of sensed and feedback signal intensities. Halfway between the "ba" and "ga" signals there are signals that derive their intensity from both the sensed representation and feedback representation. The strongest one of these signals may become the new winner and the official auditory percept. Obviously this signal represents a syllable that is not "ba" or "ga", instead it will be something in between, like "da" and thus the McGurk effect has emerged.

Thus it can be seen that the proposed machine architecture can support visually assisted auditory perception and may also give rise to McGurk type effects.

CHAPTER 14

MODELS FOR MOTOR FUNCTIONS

Motor Acts and Sequences

A robot interacts with the world mechanically. It moves around, learns to navigate and avoid obstacles. It manipulates real world objects according to the tasks given to it. While doing this it must also protect itself and others against damage and destruction. Therefore a cognitive robot must be able to learn, plan and imagine the required mechanical operations, responses and strategies. The robot must also learn to realize and control motor actions by its inner imagery and inner speech. The robot must also obey external verbal commands, a function that can be realized if the requirement of control by inner speech is met. In the following I will outline how these requirements could be satisfied within the framework of the proposed machine architecture.

Mechanical operations consist of motor acts that are executed by effectors. It is assumed that each effector is able to execute a limited number of elementary actions that cannot be divided into smaller acts. These acts are called here motor primitives and could consist of increments of twisting, turning, reaching, etc., operations. The effectors that execute these primitives are commanded by neural signals.

Distributed signal representation is also suitable for motor primitives. Distributed motor signal representation would describe which motor primitive should be executed and with what kind of force and speed.

Motor act sequences are sequences of motor primitives. For instance walking would be a repeating sequence of motor primitives that amount to moving one foot ahead of another.

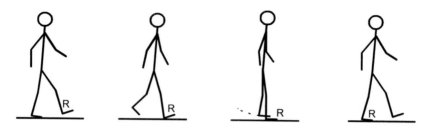

Fig. 14.1. Walking sequence

The walking sequence could be described as follows: Step one; move your right foot ahead, step two; move your weight on your right foot, step three; swing your left foot forward, etc. More generally a motor sequence like that of walking may be defined exactly by the order and timing of individual motor primitives. This information can be given with a sequence timing chart.

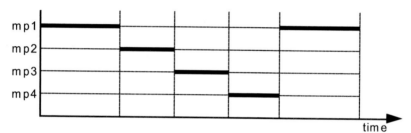

Fig. 14.2. A motor primitive sequence timing chart

A sequence timing chart lists all required motor primitives (mp1, mp2, ...) and indicates their timing and duration. Several motor primitives may be executed simultaneously. Washing machines, cuckoo clocks, toy robots, etc., operate quite nicely with this approach. However, a motor primitive sequence as defined by a sequence timing chart like the one above has a serious drawback, it is fixed and cannot accommodate disturbances.

For instance, walking is not really that simple. Balance must be maintained also on uneven surfaces, the speed of execution of individual primitives must be adjusted to suit the changing conditions, a number of correctional motor primitives may have to be included in the sequence. Adjustments must be made for starting, turning, climbing, descending and stopping. Where to go, when to turn — the system must plan these a little bit in advance and see that these are being executed as planned. Thus actual motor acts cannot be based on fixed motor primitive sequences and sequence timing charts. Instead, a sequence timing chart should be seen as a set of alternative and competitive motor primitives. In this way flexible action that is adjusted to suit the instantaneous requirements can be achieved. Many motor acts involve the serial combination of different motor act sequences. Therefore fluent branching must be possible in order to achieve these. Feedback may be used to select proper motor primitives instead of others so that the difference between the executed action and the desired action will be minimized.

Controlling Motor Acts

The controlled execution of meaningful motor action calls for the ability to initiate motor primitives at proper times and order. In this model motor primitive acts are taken as hard-wired responses to respective motor neuron excitations. Thus motor primitive action will be controlled by the excitation of respective motor neurons.

A practical robotic system will necessarily have a very large set of motor primitives. The machine should be able to imagine and plan sequences of motor acts using these motor primitives and execute these by directing proper commands to the motor neurons. It is clear that a system can be designed and wired rather easily in such a way that the motor neurons can command the actual effectors. However, it is not so clear how the motor neurons should be connected to the rest of the cognitive machine so that the planning and control of motor acts by inner imagery could take place.

Consider a video recorder remote control. There are separate buttons for the functions of play, forward, rewind, etc. You can consider these functions as the motor primitives of the video recorder system and the buttons as the excitation inputs for the respective motor neurons (even though obviously in the video recorder there are none). If you want to play the tape you will have to push the button that is marked as "play" and so forth for the other functions. Likewise a cognitive machine must excite proper motor neurons in order to initiate desired motor primitives and their combinations. However, there is a problem here. We can see and perceive the buttons and their markings on a remote control but a cognitive machine as described here (or the human brain for that matter) is not able to perceive its innards, the actual excitation input points of the motor neurons. What is available to the cognitive system is only the inner image of the desired act. It is like having a strange remote control with buttons having no markings. You know what you want but you don't know which button to push. In the cognitive machine there is no inherent connection between the inner image of an action and the motor neurons that could cause the desired action. This I call the missing link problem. This missing link problem is depicted in fig. 14.3.

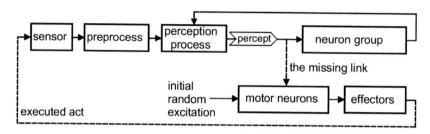

Fig. 14.3. The missing link problem

To bridge this gap the system must create an association between the inner image of the intended act and the excitation of those motor neurons that lead to the execution of this imagined act. To see how this could be accomplished we can again consider the remote control analogy. In order to learn to use a remote control with no markings on it one must push each button and note and memorize the effect. Likewise a cognitive machine may at first randomly excite motor neurons and perceive the resulting action. The percept of this action can now be associated with the excited motor neuron by the basic learning mechanism of the associative neuron.

As discussed before, within the cognitive machine model an imagined action is perceived in the same way as a sensory percept, therefore the established link between the percept and the respective motor neuron will be valid for imagined action, too. Therefore within the framework of the cognitive machine model *motor action can be readily controlled by inner imagery*. This can be done without any need to perceive the innards of the neural network. For the machine the learned pathway from imagery to action will be transparent, or if you insist, "subconscious". This pathway will also allow the instant execution of new sequences of acts. What can be imagined can be readily executed, albeit within the mechanical system restrictions. In this way non-preprogrammed motor responses become possible. Motor acts can be executed as soon as the environment evokes an inner image of them.

Now we can consider feedback control in more detail. Suppose that the cognitive machine is set to execute a certain motor sequence. According to the above inner imagery of the sequence to be executed is enough to initiate the execution of the action. Remember, inner imagery is perceived as visual percepts. These visual percept signals are associatively coupled via the learned pathway to the motor neuron group where they activate respective motor primitives as described before. The actually executed motor action can be perceived via visual and other sensors. These percept signals constitute the feedback signal to the system. Visual perception provides now the percept of the imagined and intended action as feedback from the visual neuron groups and the sensed percept provides a measure of the actual execution. The possible difference between these two provides the execution error. How much should the execution be corrected? The execution error itself evokes compensating motor primitives and the action will be corrected. Match/mismatch signals are derived from the perception process and they indicate the successful/not successful state of execution. The principle of feedback controlled execution of motor action is depicted in fig. 14.4.

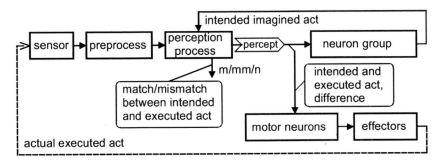

Fig. 14.4. Feedback controlled execution of motor action

So, motor action should not be generally seen as fixed execution of motor primitive sequence timing charts, instead it should be seen as inner imagery controlled action with feedback correction having many degrees of freedom.

Certain motor skills involve the execution of motor primitives in strict order, here the sequence timing chart model can be applied successfully. In this case the sequence is not necessarily optimally stored in the visual neuron groups as the evocation of the sequence could then only take place via the visual percept neurons. This route would be susceptible to disturbance from visual sensors and also non-related evocations from the visual association neurons initiated by other sensory modules. Therefore it would be more beneficial to store this kind of sequence directly in the motor neuron groups while the action could still be initiated by inner imagery. Thereafter the motor neuron group would execute the action and the rest of the system would only perceive the outcome. The execution would be automatic in the sense that the action would not be preceded by driving mental imagery, nor would there be available any intended action feedback to visual perception/response module perception process and the execution would not be corrected. This mode of action would obviously correspond to highly rehearsed and automated human skill execution, where the performer sometimes reports being like an outsider who only observes what happens.

Mirror Neurons and Imitation

Imitation is a highly useful method in learning and skill acquisition. While motor skills, complex sequences of motor primitives can be learned by the tedious trial and error method, imitation would offer direct access to skills that others have already acquired.

As an example let's consider the imitation of facial expressions. To be able to imitate the seen facial expressions of others the system must know which motor commands will produce similar expressions to those seen. This problem is exactly the missing link problem of the previous paragraph and can be solved by observing the results of random excitation of motor neurons as I have described earlier. In this case the machine might produce various facial expression primitives on its own and perceive these via a mirror. Then these percepts could be associated to the respective motor neuron commands.

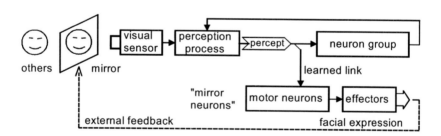

Fig. 14.5. Learning to imitate with the help of a mirror

The learning of the link between percepts and motor neuron commands is the first step towards the ability to imitate. However, this step is not sufficient alone. The additional requirement is that the observation of others to be imitated produces similar percepts to the observation of the system self producing the same primitives, otherwise the learned link between percepts and motor commands is not valid. Obviously this condition can be met with the mirror arrangement here for visual perception and without any helping devices for auditory perception and imitation.

In these cases there exists a direct pathway from perception to motor neurons responsible for the corresponding motor primitives. The activation pattern of these motor neurons will now follow or "mirror" the perceived action and the perceived action will readily evoke the respective motor output if the motor neuron output thresholds are low enough. Now imitation is possible or even inevitable. Due to this mirroring action these motor neurons can be called mirror neurons.

Little children seem to learn to imitate facial expressions without the help of mirrors. Is this machine imitation scheme therefore different from that used by the human brain? This is not necessarily so. When adults see an infant to produce something like a smile they will imitate that ("look at her, she is smiling") and this will act as the mirror to the infant.

In many cases imitation is done only after the completion of the observed action. In that case the observed action must be temporarily memorized and mentally replayed so that delayed execution of the action via motor neuron excitation becomes possible. The feedback loop of the corresponding sensory perception/response module and the neuron groups within it can provide this kind of rehearsal memory loop.

Sometimes the action to be imitated does not give rise to percepts that have direct connections to motor primitives. In that case imitation may still be possible indirectly if the percepts are able to evoke inner representations that have this required connection.

Why should the system imitate anything? Firstly, if the motor neuron output thresholds are low, acted out imitation is inevitable. Secondly, delayed imitation and its observation leads to match condition and respective match pleasure.

CHAPTER 15

MACHINE EMOTIONS

System Reactions

In chapter 6 I outlined a system reaction theory of emotions. Here I will discuss how this theory could be implemented within the artificial cognitive system framework. According to the system reaction theory emotions are combinations of basic system reactions with eventual cognitive evaluation. Likewise machine emotions are combinations of basic machine system reactions that are functionally similar to their biological counterparts. Thus an artificial emotional situation involves also a triggering event, basic system reactions to this and percept representations for these system reactions so that these can be associatively incorporated in episodic memories as well as named.

Basic system reactions are responses to elementary sensations. In a biological system elementary sensations originate from the senses of taste, smell, pain, pleasure and match/mismatch/novelty detectors.

Even though an artificial cognitive agent or robot may not have the need to digest something and would not therefore need to smell and taste things, an abstracted sense of good and bad, to accept or reject, would still be needed. This can be provided by having artificial inputs for good/bad system reaction initiation.

Likewise pain and pleasure system reaction initiation may be based on artificial inputs. However, in robotic applications sensors for physical damage may and should be used as pain sensors. These inputs can then also be used to punish and reward the system. It should be noted here that even though the machine pain and pleasure input stimuli may not necessarily be similar to those that give pain or pleasure to us humans, the system reactions that they evoke are intended to be similar. Pain and pleasure are not properties of their evoking stimuli; instead they are a system reaction evoked by the stimuli. Therefore the use of artificial evoking stimuli does not exclude the possibility of machine system reactions that are very similar to the biological system reactions of pain and pleasure .

The system reactions of an artificial system may not be as varied as those found in biological systems. The basic reactions may still be similar like the direction of action, execution force and speed, the effects on attention and arousal caused by varying neural signal processing thresholds and signal intensities. Table 15.1 gives a list of basic system reactions and their possible realization in a robotic system.

Table 15.1. Elementary sensation, related system reaction and main area of realization

Elementary sensation	Related system reaction	Main realization
Good; taste, smell	Accept, approach	Motor act direction
Bad; taste, smell	Reject, withdraw	Motor act direction
Pain; self-inflicted	Withdraw, discontinue	Threshold control & motor
Pain; due to ext. agent	Aggression (retaliation)	Motor drive level, direction
Pain; overpowering	Submission	Thresholds, motor drive level
Pleasure	Sustain, approach	Threshold control, direction
Match	Sustain attention	Threshold control
Mismatch	Refocus attention	Threshold control
Novelty	Focus attention	Threshold control

The connection of system reactions to the overall cognitive system architecture was given in fig. 13.1 in a very general way. Now a more detailed description of the integration of elementary sensory process, system reaction control and system reaction perception process into the machine architecture is needed, this is outlined in fig. 15.1.

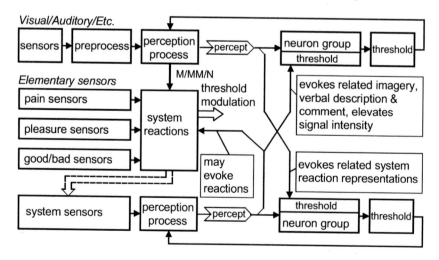

Fig. 15.1. System block diagram for the connection of system reactions

Sensory signals from elementary sensors and match/mismatch/ novelty detectors initiate system reactions. These are perceived by the system sensory modality perception/response module that creates a distributed signal representation of the perceived reaction. These system reaction percept signals are connected back to the system reaction control circuits where they are able to evoke the represented system reaction there. The system reaction percepts are also broadcast to other sensory modality perception/response modules. It should be noted that the broadcast system reaction percepts are not the only way in which the other perception/response modules may become aware of ongoing system reactions. Some system reactions may interfere directly with other ongoing activities, they may disrupt the execution of motor routines, etc. Threshold modulation is one of the system reactions, this will also interfere directly with inner attention and affect ongoing perception/response processes.

The triggering event that causes elementary sensory signals may also be perceived by visual, auditory, touch, etc., sensors. These percepts are associated with the system reaction percepts and the other way round, system reaction percepts are associated with the visual, auditory, etc., percepts. Thereafter the visual, auditory, etc., percepts of triggering events, real or imagined ones, are able to evoke representations of the associated elementary reactions, that is, the "emotional soundtrack".

The effects of the system reaction percepts on the other sensory modality perception/response modules are many. For instance, if the visual sensory modality detects many entities among which one has associated emotional significance, perhaps pain, the corresponding system reaction representation is evoked at the system sensory module and this percept will then intensify the representation of the pain-connected entity at the visual sensory module. In this way a binding loop is formed, the pain-connected entity will win other representations and become the focus of attention. The system reaction percept may also evoke an "explanation" to the elementary sensation; a representation of a possible triggering event, guided by and bound from actual visual and auditory cues. This would be an answer to questions like "what hurts", "what hit me", etc. The system reaction percept may also evoke a verbal comment like "this is good", "it hurts", etc.

The signal intensity of distributed signal representations determine which representations will pass the various thresholds and will thus become the focus of operation; the focus of attention. Thus there are two ways to affect attention, namely the control of signal intensity and the control of thresholds. Signal intensity control is a good choice if the desired distributed signal representations are known, otherwise general threshold control must be used.

Many system reactions involve the control of attention. This is in many cases achieved by threshold control due to the above reasons. Pain and pleasure reactions especially try to control attention. Pleasure tries to sustain ongoing pleasure-producing activity and suppress unrelated activities. Ongoing activity can be sustained by elevating both the activity signal levels and thresholds. Thus any unrelated activity would be effectively cut off and attention shift would be difficult. Pain tries to focus attention to the damage and discontinue any ongoing activity. Any ongoing activities can be cut off by elevating thresholds strongly. However, thresholds cannot be kept high for extended periods as this would prevent all activity and no remedial acts could be initiated. Therefore periodic up/down modulation of thresholds could be used. This would achieve the goal in quite a disruptive way and would also enforce periods of "rest" for "recuperation".

Emotions in the Machine

I am proposing here that machine emotions are interactive combinations of system reactions that determine the style of cognitive processes and actual responses. Some examples are considered here.

Fear is expectation of pain; fear makes the sufferer to try to escape the situation. In the cognitive machine "fear" involves the evoked representation of pain, focussing of attention to the fear-causing entity, reversed direction of motion and elevated motor drive levels. The machine, just like humans may sometimes solve a fearful situation by basic stimulus–response action; by simply backing off and escaping. In that case "fear" would only appear as a trivial transient event. However, in many cases escaping is not possible and the agent, robot or human, must force its way ahead in spite of the fear. In these cases conflicting motor commands exist, ones that try to make the agent move forward and ones that try to make the it to back off. The more the agent moves forward towards the cause of the fear the stronger the opposing force will be and the more intense must the forward command be made by internal or external motivation. Now the external symptoms associated to fear may become evident; those that appear as reluctance, hesitation and trembling caused by conflicting forward/reverse commands.

The artificial cognitive system is also able to imagine fearful situations, to have a flow of inner representations of these, both visual and verbal. In these cases also the response is imagined without any need for actual motor action. The abstract conflict of approach and withdrawal may still cause fear-related symptoms. Elevated signal levels may be present, therefore motor response thresholds might be occasionally exceeded and actual motor trembling might very well occur; the system would have physical stress-like symptoms caused by sole inner machine thought processes.

Desire makes the agent long for something. In the cognitive machine the basic system reactions are pleasure, evoked by the desired entity, attention to this entity and forward direction of action. Why would a machine desire something? We could link pleasure to certain objects and make the machine, a robot, to pick up these. For instance we could make the robot to pick up and collect trash or other items. This by itself is again trivial, but what happens when the robot is not able to move around? Images of trash will pop up in the cognitive system trying to initiate collection routines. However, these routines are now blocked and mismatch-displeasure follows. There will be a conflicting situation between the pleasure linked to the images of trash picking and the displeasure evoked by the mismatch between the planned execution and actual situation. This is typical of desire.

Anger is characterized by aggressive behavior against an object. Anger may be triggered by physical assault or an object's failure to meet the agent's expectations. Anger can be seen as an attempted brute-force solution to a problem. In the cognitive machine anger-like responses may arise when sensory pain or mismatch displeasure is inflicted repeatedly so that a solution by withdrawal is not possible. This will elevate motor drive levels so that any response will be executed forcefully. In humans anger is often accompanied by attempted physical retaliation often involving learned motor routines. Even though anger has a survival function I am not sure if "anger" especially with learned, perhaps lethal, retaliation motor routines would be useful for robots generally. In that case it would remain for the designer to ensure that anger-like behavior would be suppressed and no retaliation responses could be learned and executed.

So, it seems possible to reproduce the functional reactions and responses of various emotions in the machine. Also the tagging of objects and memories with emotional significance is possible. Emotionally significant events can be memorized instantly. Machine emotions are subject to cognitive evaluation in the same way as their biological counterparts. Machine emotions can thus be made functionally and as processes similar to human emotions even though the spectrum of machine emotions may not necessarily be as wide. The question remains: Are machine emotions felt, are they accompanied by first person experience? This question is related to that of consciousness and especially to the conditions on which a system can be aware of its own system and cognitive states and how these would manifest themselves to the system.

Machine Volition and Motivation

Volition, the act of willing is easily understood as a property that is owned by conscious beings only. Free will seems to manifest itself in the freedom of choice; to choose between actions, to choose between thoughts. Predetermined sequences of acts are not a result of any selection process, therefore free will would necessarily seem to involve some non-deterministic process.

A digital computer is definitely without free will. A computer executes exactly whatever sequences of commands are programmed, whatever conditional branches are provided. The computer may execute commands of the type: IF (*condition A*) THEN (*result B*) ELSE (*result C*). Here the computer makes a decision; it selects a response that depends on a given condition. This is decision making, but the process is completely deterministic; the computer itself does not have a choice. The computer does not even understand what it is doing as it does not operate with meanings.

The artificial cognitive machine is not governed by any programs and therefore will not execute any preprogrammed decision commands like the IF–THEN ones. Nevertheless, the cognitive machine must be able to make decisions if it is to do anything useful at all. Useful tasks may be given to the machine to be executed whenever needed or the opportunity arises. The machine must also take care of itself, it must see that it has enough stored energy, it must not damage itself, and it must avoid dangerous situations. In order to manage these the machine must choose between options, what to do or not to do and when. Superficially this would seem to be a case for the IF (*condition A*) THEN (*result B*) ELSE (*result C*) command structure. However, apart from trivial cases, how do we foresee and program all possible *conditions A* and suitable results for these? We cannot and therefore the cognitive machine must cope on its own. This it can do because it does not rely on preprogrammed commands, instead it uses the perception process, significance evaluation and the flow of inner imagery and inner speech as the basis of its decisions.

The machine must make a decision whenever inner imagery for mutually exclusive actions is evoked; which one if any of these is to be executed. Who or what in this machine makes the decision? In the artificial cognitive machine model there is no specific decision making machinery, no box labeled as the "machine self" or supervisor. Instead the decision making is a distributed process involving the focussing of inner and sensory attention.

Attention in the machine operates with signal strengths and thresholds. These in turn are controlled by match/mismatch/novelty conditions and emotional significance; good/bad and pain/pleasure. Here the basic system reactions and emotional reactions have their effects on what will be executed and in which way as was discussed before. Thus decision making in the machine is a process that is controlled by reasoning related match/mismatch conditions and machine emotions related significance. Obviously this process allows much more variety and adaptability than the program based IF–THEN–ELSE decision structure. A computer does not consider alternatives, it blindly and instantaneously executes the response that is dictated by the detected condition. The cognitive machine is different; it may entertain various alternatives at the focus of attention for a while and only then one of the alternatives will be selected when its intensity rises over an execution threshold. In addition the cognitive architecture allows the introspection of the process; the machine may have inner linguistic comments about the situation like "what is this", "what should I do" or "I feel like doing this", etc., and also "I did this because this is better".

Does the machine have free will then? Are these decision processes non-deterministic? The decisions are based on the machine's needs, preferences and "emotional state", which can vary from time to time, even during the act of decision making. Also the choices that are offered by the environment vary and may not always be perceived in the same way. The distributed signal representation method itself allows variation in the associative process. Therefore superficially similar cases may not necessarily lead to similar outcomes. The decision process may in principle be deterministic but in practice it may not be so.

However, in the section "Consciousness and Free Will" (chapter 8) I argued that conscious free will cannot exist. Thus it would be bad logic to use the concept of "free will" here. The machine's decisions are based on its needs and preferences, therefore instead of any questionable "free will" they reflect something that could be called the machine's own will.

Generally speaking, attention in the machine is focussed on what is good or pleasant and actions that produce these tend to be imagined and executed, while the bad or unpleasant tends to be avoided. "Machine will" can be used as a name for this process and tendency.

Machine will may be guided by rewards and punishments and in this way the machine can be motivated to do something. A reward for doing something associates pleasure with the intended task. According to the defined effects of pleasure this kind of action will be sustained and the task will be executed. In practice the emotional significance of pleasure translates into elevated signal intensities and these in turn will secure that the related representations keep on winning at threshold circuits and thus the machine will keep on executing the pleasure-producing task.

A punishment works in the same way. Here pain is associated with an act that should not be executed. According to the defined effects of pain this kind of action will indeed be discontinued. Pain will disrupt attention to the unpleasant task, that is, the threshold levels are periodically raised and consequently the task is discontinued because the required representations are no longer passing threshold circuits.

Thus the use of rewards to make the machine do something and punishments to make the machine not to do something seems easy enough and is readily achieved via the defined system reactions of pleasure and pain. But then we know that in real life the roles of reward and punishment can be inverted, rewards can be used to have somebody not do something and punishments may be used to make somebody do something, now can this be implemented here at all?

The real action of pain and pleasure is not inverted here after all. When we punish somebody in order to make him do something then actually we associate pain with all alternative actions so that they will be discontinued causing the desired action to remain as the only pain-free alternative. Likewise we can reward alternative action that becomes then executed instead of the undesired action. Now again these functions can be achieved within the cognitive machine architecture via the defined system reactions of pain and pleasure, no modifications are needed.

Let's suppose now that the machine is executing some unpleasant task. Obviously this task is on the verge of being discontinued, but what happens next? If the threshold levels are heightened then it depends on the strength of other available representations, other activities, which one will win. If nothing else is available, then it is possible that the unpleasant task will be continued after all. However, if more pleasant activities are available then one of them will win and will be executed.

Likewise, the signal intensity of representations can be used to indicate the importance of the task. Virtual task stacks can thus be created so that the task popping up as the most important will be executed at any given time.

CHAPTER 16

LANGUAGE AND INNER SPEECH

The Implementation of the Multimodal Model of Language

The multimodal model of language that I have outlined earlier assumes a representation plane for each sensory modality. These planes learn and acquire their representations via the perception process. Each plane contains representations of its own kind; the visual plane handles visual objects, the auditory plane handles sound patterns, etc. Representations within a plane can be bound into groups and these in turn can be linked to each other. These representations can also be vertically connected to the representations in the other planes. The multimodal model of language solves both the horizontal and vertical grounding problems for the meanings of concrete and abstract words.

The cognitive machine architecture provides directly the required representation planes for each sensory modality. It also provides the necessary horizontal and vertical connections. Distributed representations can be manipulated in the ways demanded by the multimodal model of language. In the following I will outline how the cognitive machine architecture enables the learning of words and the acquisition of vocabulary and syntax, the basic elements of language. My aim here is to outline a machine that is able to think and communicate by language and speech.

Speech Acquisition

Speech acquisition begins with the imitation of sounds and words. The neuron groups within each sensory modality are able to learn sequences and use these as entities as they were specified so. Thus the auditory modality has the build-in ability to learn and handle limited length auditory signal patterns as entities, words that can be associatively connected to each other and other sensory modality percepts.

It is assumed here that the cognitive machine has an audio synthesizer as an output device. This synthesizer is supposed to be able to produce the sound primitives that are needed for the production of complete words. This synthesizer is supposed to be associatively coupled to the auditory percept point so that auditory percepts, heard or imagined, may excite it. Now the machine will be able to learn to utter words via imitation. The process of learning by imitation is illustrated by fig. 16.1.

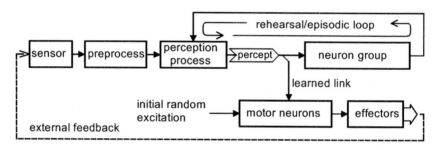

Fig. 16.1. How the system learns to imitate sounds and words

The imitation of an instantaneous sound involves the reproduction of the perceived sound by the sound synthesizer or audio effector. It is assumed that the sound synthesizer is designed to be able to produce a large variety of simple sounds, sound primitives, and each of these primitives can be produced by the excitation of a respective control signal. Thus the imitation of a perceived sound calls for an associative connection between the sound percept and the motor neurons that cause the production of a similar sound. These kinds of exact connections do not have to exist initially, instead they can be learned associatively. Let's assume that the sound producing motor neurons were initially excited randomly so that each sound primitive would be produced in turn. These sounds would be coupled to the auditory sensors externally and be perceived by the system. Thereafter these sound percepts would become associated to the motor neuron excitations that were producing them. The link between perceived sounds and respective control signals would thus be formed and later on any perceived sound would associatively excite the respective control signals and these in turn would cause output if the thresholds were low enough. We can see again the "mirror neuron" action here and via this mechanism the imitation of sound primitives becomes possible.

Temporal sound patterns are sequences of instantaneous sounds. Therefore the machine must also be able to learn temporal sequences and reproduce these. The neuron groups have this ability as they are so configured with the help of built-in short-term memories.

Complete words should be imitated only after they have been completely heard. This can be achieved if an echoic or rehearsal loop that sustains the heard word is included in the system. Later on this sustained representation can be used to excite the required control signals so that the word could be imitated. In fig. 16.1. the inner feedback loop could act as the necessary rehearsal loop.

Learning to Read

It might be thought that reading is a simple matter of recognizing written letters and words. But then again reading involves also the evocation of meaning, inner imagery, background information, emotions, etc., and therefore may not be that simple at all. So, what is the truth here? In what follows I will propose that the reading process can be quite simple in a proper cognitive system. If the cognitive system has machinery for the understanding of spoken words and oral narration then the understanding of text can be facilitated by the same machinery so that very little additional burden remains for the reading process. It really does not matter via which sensory modality the words are received.

Within the proposed cognitive architecture spoken language is processed by the auditory sensory modality and related motor neuron groups in ways that will be discussed in more detail later on. In order to utilize these language processing capacities the visually perceived letters and words should be forwarded to the auditory module where they should evoke the respective auditory representations. This can be done readily via the associative cross-connections that are inherent in the proposed cognitive architecture.

Thus the first step towards the ability to read is the learning of alphabet characters by associating these with the corresponding auditory phonemes. This can be done by ostension; by repeatedly pinpointing the character and simultaneously pronouncing the corresponding sound. When the training is completed the system will have the necessary associations for the full alphabet. Due to the two-way nature of the cross-connections the system will now be able to evoke the corresponding phoneme for a visually perceived character and also to evoke the visual forms for heard phonemes. This process is illustrated by the fig. 16.2.

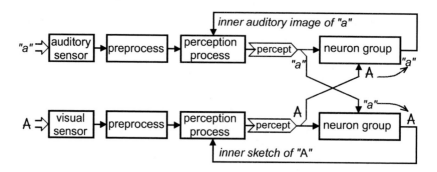

Fig. 16.2. The association of a visual character to the corresponding
phoneme

In fig. 16.2 the visually perceived character (A) is associated to
the corresponding phoneme ("a") at the auditory neuron groups via
repeated coincidences of these percepts. Likewise the phoneme ("a") is
associated to the character (A) at the visual neuron groups.

Reading involves the scanning of the letters of a word
sequentially from left to right so that each letter becomes a percept in
turn. The scanning is facilitated by the small fovea and the changing of
gaze direction by rotating the visual sensors. The scanning process is
actually a motor sequence that must also be learned. This sequence can
be taught by repeatedly pointing out the letters of words in correct
order. Thereafter the scanning sequence will be stored in the gaze
direction motor neuron group and can be evoked by an associated cue,
like the perception of letters.

Words are processed as serial streams of sounds, as sequences
of phonemes with the mechanisms that I have described earlier.
Therefore, when a written word is visually scanned, the perceived
letters will evoke a sequence of phonemes which, by the nature of the
process, will be treated by the auditory module as a word and it will
evoke all the associations that a heard word would do.

In phonetic languages like Finnish there is a one-to-one
correspondence between a phoneme and its character. Each character
depicts only one phoneme, all phonemes are written and all characters
are pronounced. Therefore for these languages there is not much more
to it, the described process is sufficient for the evocation of auditory
words by the corresponding characters. However, many languages like
English, French, etc., are not phonetic. Here more complicated learning
is needed. One-to-one correspondences between characters and
phonemes do not usually exist, instead associations must be made
between groups of phonemes and characters.

Eventually perfect letter-by-letter scanning is no longer needed as partially perceived words can be associatively completed by expectation, connections to the sentence and background information. In this way we come to the problem of word meaning.

Vertical Grounding of Word Meaning

In the multimodal model of language the basic meanings of the words are grounded both vertically and horizontally; to percepts from various sensory modalities and to other words and sentences. The cognitive architecture allows directly the cross-association of percepts from various sensory modalities to the special sound pattern percepts, words. These cross-associations can be achieved by ostension, pinpointing the entities to be named, be it a visually perceived object or action or a percept from any other sensory modality.

However, there are entities like "small", "large", "round", etc., that cannot be pointed out separately as they only appear as properties of other entities. Correlative learning allows also the learning of these kinds of properties and is used in this cognitive model. The correlative learning mode calls for repeated coincidences of the percepts to be associated. Now several different objects that share the property to be named can be used as examples and eventually only the desired property gets associated with the given name.

According to the neural model that is used here two representations are associated together by repeated temporal coincidence. However, in practice a large number of representations coincide all the time and only few if any of those connections may actually be desired at that moment. Therefore another selective mechanism is needed to allow only the intended associations to take place. This mechanism is attention. Ostension is a process that focuses attention to the selected entities. This in turn manifests itself as elevated signal intensity of the signals that constitute the respective distributed signal representations. The actual synaptic association may now be allowed only between signals with elevated intensity.

Words that can acquire their basic meanings by vertical grounding are concrete as their basic meaning can be pointed out. This grounding of meaning is a two-way process; the named percepts may evoke their name and vice versa, the name may evoke the respective inner representation for the named entity, fig. 16.3. The actual association of percepts with words takes place at the auditory neuron groups, and the association of words with percepts at the neuron groups of the other sensory modality loops, respectively.

Fig. 16.3 depicts one-to-one correspondence between percepts of entities and respective words. However, vertical grounding can be taken beyond this trivial "one entity — one name" concept. For instance, category names designate larger sets of entities. Thus each member of the category will have two names associated to it, its specific name and the category name.

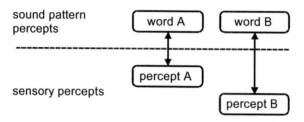

Fig. 16.3. Vertically grounded words derive their meaning from association to a corresponding sensory percept

Let's consider the category "color" as an example. The human eye utilizes three primary colors and their combinations to represent all the colors that we can see. Likewise three primary colors, red, green and blue are also used in color television. If we assume that these three primary colors were also used here, then the names of these colors "red", "green" and "blue" would be associated with the percepts of these colors and vice versa. But how about the other colors like yellow? The percept of yellow arises from the additive combination of red and green. Thus the name "yellow" would be associated to the simultaneous appearance of red and green.

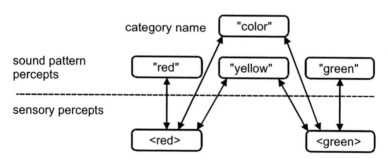

Fig. 16.4. The association of color names and the category name

The category name "color" applies to each and every possible color. Must we now painstakingly associate the word "color" to every perceivable color so that a proper category would arise? No. It suffices that the name "color" is associated with just the primary colors red, green and blue. Every other color is a combination of these and the constituting primary colors can therefore evoke the category name "color". This will also apply to novel colors, those that the system has not seen before and does not even know their names, fig. 16.4.

We can see that in fig. 16.4 each percept may have a number of associative connections to various words. What happens now if, for instance, the color <red> were perceived? Which word would be evoked, "red", "yellow" or "color"? Is there an unequivocal answer or do we have another fine mess here? In principle the word with the strongest evocation would be selected. The word "yellow" would seem to have only partial evocation, but "red" and "color" would seem to have equal evocation strengths. However, it should be understood that in practice each word would also have associative connections to elsewhere in the system. Therefore the actually evoked word would be determined by the instantaneous contextual state of the system.

In the above example the word "yellow" was determined by the simultaneous presence of multiple properties, namely <red> and <green>. This is an example of a more general case where a name would be associated with a specific combination of the constituting property percepts, fig. 16.5.

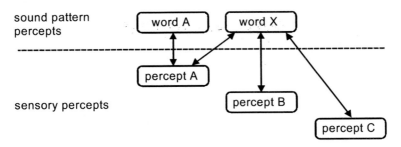

Fig. 16.5. Naming an entity with multiple property percepts

In fig. 16.5 the word "X" is associated to the percepts A, B and C. These percepts must appear simultaneously in order to be able to evoke the word "X". However, the word "X" is able to evoke all these percepts simultaneously.

On the other hand the percepts A, B and C may already have corresponding names appropriate to themselves and these names may evoke their respective percepts as inner imagery. This opens up the possibility of *learning by verbal description*. Instead of showing an object with the properties A, B and C we can verbally list these properties and state the name of the corresponding object, the word "X". Now in the machine the inner representations for the percepts A, B and

C are evoked. The combination of these will then be associated to the entity name "X". Learning will take place if the process is repeated. In this way the machine can learn entities that it has not actually seen before and may recognize these if they eventually show up. Due to the two-way association the name "X" can evoke the inner representations for its properties A, B and C. This will allow inquiries about X like does X have A? Let's take the dollar bill as an example. A naïve description of the dollar bill is: "Small, square, green". Suppose that the word "dollar" has been associated to these properties. What happens if we ask what the color of the dollar is? The word "dollar" will evoke the inner representations for <small>, <square> and <green>. The word "color" will evoke color representations, thus also the representation for <green>. The representation <green> will be most strongly evoked and will therefore be able to evoke the corresponding word "green". In this way the correct answer will be produced by evoked inner imagery, fig. 16.6. Likewise we could have asked what is the shape or size of the dollar.

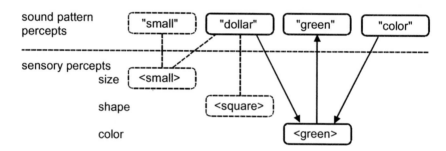

Fig. 16.6. "What is the color of the dollar bill"

Do these principles of vertical grounding really work as proposed or is this only wild speculation? Computer simulation is one way to find out. I have written an artificial cognitive system simulation program "Cognitive Processor"[41], which incorporated vertically grounded language according to the above principles. This program was written in Visual Basic, utilized distributed signal representations, ran on an ordinary PC and took its input from keyboard and live video camera with limited visual resolution.

Fig. 16.7 shows the user interface of the program. Here "Inner Image" depicts the shape features of the visually perceived or imagined object. "Color/Size" depicts the detected color and size of the object (colors: up red, middle green, blue bottom, sizes: up large, middle medium/neutral, bottom small). "Input Word" depicts the instantaneous input word from the keyboard. "Inner Word" depicts the instantaneous inner word at the word perception point. "M/MM/N2" depict the instantaneous inner match/mismatch/novelty condition. Each horizontal column depicts one time point, latest time point being at the bottom.

In the dollar bill example the concepts of size, shape and color had been taught to the system via correlative association, by showing and naming suitable example figures. Then a small green square was shown to the camera and was named "dollar". Thereafter the visual input was blanked and the question "dollar what color" was typed in. The subsequent situation is depicted in fig. 16.7. Here the word "dollar" in the sentence "dollar what color" evokes an inner image of small green square and the constituting feature percepts are sustained by the respective perception/response loops. The word "color" primes every color at the color sensory modality loop, therefore the active percept <green> is amplified and is thus able to evoke the response word "green". It can also be seen that the response word "green" evokes inner match condition against the imagined color green. In this way vertical grounding to perceived or imagined imagery can be used to answer questions about seen or imagined entities.

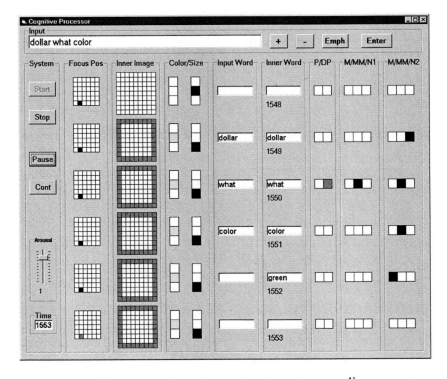

Fig. 16.7. Artificial cognitive system simulation program[41]; Answering questions by vertical grounding to sensory percepts and inner imagery

What kinds of words can be grounded vertically in this way? In general, whatever signals can be generated and broadcast to the linguistic system can be labeled. These signals can depict objects, properties, relations, actions, and system reactions. In this way the basic meaning for nouns, adjectives, verbs, etc., can be grounded. However, it should be noted here that within the cognitive architecture the meanings of words will not be depicted by a single association only, instead all words, also concrete ones, will acquire a network of vertical and horizontal associations that also convey meaning.

Vertically Grounded Sentences

A sentence is a string of words that describes events, entities and their relationships. In the multimodal model of language the entities and their relationships are the percepts, either of external world or imagined within the various sensory modalities. A sentence is therefore a verbal description of the instantaneous perceptual state of the system. How does this kind of a description arise? Let's consider as a simple example a case where the cognitive system is "hungry". "Hunger" is perceived via a suitable internal sensor and this sensory percept may be associated with the word "hungry". It is also assumed here that a number of other words have been taught to the system.

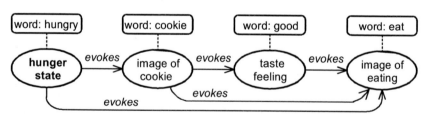

Fig. 16.8. Production of vertically grounded sentences, an example

In this example perceived <hunger> evokes the word "hungry". It also evokes imagery of edibles, for instance <cookie>. This in turn evokes its attribute <good>. The imagery of edibles evokes the imagined action <eat>.

The respective words "hungry", "cookie", "good" and "eat" may be evoked more or less simultaneously because they involve separate sensory modalities and separate neuron groups within the auditory modality. The auditory perception/response loop winner-takes-all threshold passes only the strongest word representation of these and forwards it via the feedback loop to the auditory percept point. Neuron outputs are provided with a decay function so that the signal intensity of the winning representation will eventually level out allowing the next representation to win in turn. In this way a string of words, a report of the system's instantaneous percepts; needs, imaginations, proposed actions, is generated: "Hungry... cookie... good... eat". The word order of this string is determined by the relative significance of each evocation. This vertically grounded thought sentence will in turn evoke further associations in other sensory modalities and prime perception process. In this example the percepts of the imagined "cookie" would prime sensory modules to look for cookies.

However, strings of words like this are not real sentences in a strict grammatical sense. Each word stands here alone capturing only its respective concrete meaning. These sentences do not have proper syntax, they do not adopt the word order and structure of the speech of others. With syntax the generated strings of words could look like this: "I am hungry. Cookies are good. I feel like eating cookies". A system that generates vertically grounded "speech" knows what it is talking about, but this "speech" is limited to very concrete matters and lacks word-to-word connections. Vertically evoked words are not able to evoke other words directly, the only access to other words is via evoked inner imagery and other inner representations as described before. The system cannot learn word-to-word associations, it cannot memorize facts as connections between words.

Artificial speech that is grounded vertically only has a striking similarity to the neurological condition known as Broca's aphasia[9]. Patients suffering of this condition utter strings of words with little or no syntax at all, but their comprehension is not impaired. Broca's aphasia is related to a brain damage at the rear of the left frontal lobe, also known as Broca's area. Broca's area is thought to be involved in the production of speech, however, I might speculate that more exactly this area could be involved in the horizontal grounding of speech, the formation of word-to-word links.

Horizontal Grounding of Word Meaning

Horizontal grounding of word meaning operates on heard strings of words and complete sentences. In principle the heard words are associated together, not as serial strings like words of verse but in a parallel way.

Only entities that are present simultaneously can be associated together. Therefore short-term memories are needed to sustain previous words of a sentence so that they can be associated to later ones. Within the cognitive architecture this association would take place at the auditory perception/response module neuron groups.

Let's consider a simple example. What kind of associative connections between words could arise from the sentences "banana is fruit", "banana is yellow", "banana is not blue" and "fruit is good"?

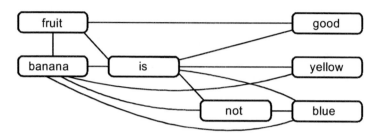

Fig. 16.9. The word associations from the example sentences

In fig. 16.9 the associative connections are indicated by interconnecting lines. The associations work both ways and may have variable strengths even though here constant strengths are considered only. We can see that the words "is" and "yellow" evoke the word "banana", as this word will be evoked most strongly by the strength of two connections. Likewise "banana is blue" will evoke the word "not".

Surprising things can be achieved by horizontal grounding. Fig. 16.10 depicts a sample output from a simulation program that processes sentences in the style of figure 16.9 without any vertical grounding of word meaning.

In this example four sample sentences are given (banana is fruit, banana is yellow, banana is not blue, fruit is good). The program has no stored vocabulary or grammar and encounters the words of these sentences for the first time. The inequality mark (>) indicates that the following sentence is input, sentences without this mark are responses generated by the program. The exclamation mark (!) indicates that associative learning is to be executed. After inputting the four sample sentences questions are typed in and the program generates responses by the associative links that it has created from the sample sentences. Strangely enough the program is able to answer to the question "what is banana" with "banana is fruit yellow not blue". It is also able to oppose the claim "banana is blue" with "banana is not blue".–The word "what" is not associated anywhere in this example and is therefore actually ignored by the program.

We can see that superficially this program works like an encyclopedia, perhaps a clever one that is able to combine information from multiple sentences and respond to questions that have arbitrary wording. However, the blind associative evocation used by this program does not guarantee correct responses every time.

Fig. 16.10. An example of horizontally grounded language

A system that generates horizontally grounded speech without any vertical grounding cannot report verbally what it perceives via other sensory modalities, it cannot report its own mental imagery, needs or other states. Moreover, the system does not really know what it is talking about, but nevertheless the speech may imitate correct grammar and even be factually correct. This situation is remarkably similar to a neurological condition known as Wernicke's aphasia[9]. Patients with this condition are able to produce fluent sentences with occasional apparent meaning and syntax but without comprehension. Obviously the patients do not properly comprehend heard speech either. Wernicke's aphasia is related to damaged area at the left hemisphere, rear of the left temporal lobe, also known as Wernicke's area. Wernicke's area is thought to be involved in the comprehension of speech. However, Wernicke's area is quite close to visual sensory cortex areas and could therefore be involved in the vertical grounding of speech, the formation of meaning by word-to-percept links. Damaged Wernicke's area would inhibit vertical grounding of meaning. In that case any fluent nonsense speech would be produced by the Broca's area, thus responsible for the horizontal grounding and word-to-word connections. Accordingly intact Wernicke's area alone and on its own would produce meaningful strings of detached words, vertically grounded "sentences".

Combined Horizontal and Vertical Grounding

Speech that is based on vertical grounding of word meaning only is like a very rudimentary running commentary on the flow of sensory percepts. The total lack of word-to-word connections prevents the generation of sentences without any sensory stimuli; pure verbal imagination is not possible. On the other hand, speech that is generated by horizontal word-to-word connections only is meaningless and lacks practical value as it cannot evoke or be evoked by inner imagery of objects and actions.

A simple example may illuminate the situation. Candy is good, this we know by sensory experience. By vertical grounding the words "candy" and "good" may be associated to the sensory percept representations. Now, without any horizontal word-to-word connections the word "candy" cannot evoke the word "good". However, via the vertical connections this can be made. The word "candy" may evoke an inner representation for the percept of candy, this in turn may evoke an inner percept for goodness due to the relationship between candy and goodness. This inner representation of goodness may now evoke the word "good", fig. 16.11. Also system reactions like those of desire may be evoked.

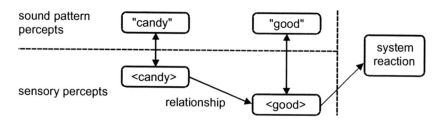

Fig. 16.11. Vertically grounded words are indirectly connected via real-world relationships

The addition of horizontal word-to-word connections allows the representation of sensed real-world relationships in speech, fig. 16.12. In this example the words "candy" and "good" become connected so that "candy" may evoke the word "good" without the help of inner images of sensory percepts. Furthermore a connection to system reactions may be established. Thus the word "candy" may by itself evoke the system reaction of desire. Additional words may be used to describe the real-world relationships.

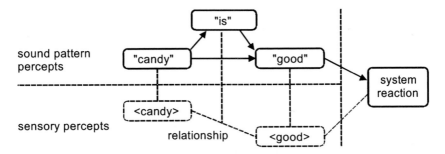

Fig. 16.12. Combined horizontal and vertical grounding

Using combined horizontal and vertical grounding we get a linguistic modality that is, in principle, able to reflect real-world entities and their relationships. This modality will allow verbal thinking in the form of structured strings of words that in turn evoke further strings of words, even without the necessity to evoke any inner imagery.

On Syntax

A typical sentence is a description or a comment of the state of affairs. In many cases it is a short story in itself, telling what is what, who did what to whom, when, how. A sentence may describe quite complex relationships between the involved entities. Moreover, the information within a sentence is to be coupled to previously given information and to the information to be subsequently given. This is a complex task and a simple sentence without a structure, consisting of strings of associatively connected words only, will not usually be able to convey the required relationships properly and ambiguity may remain. Thus we need a suitable structure, syntax, to remove ambiguity from sentences so that the intended relationships can be presented correctly. Some trivial aspects of syntax are briefly considered here with some examples.

As an example let's first consider the following sentences: "Cats eat mice" and "Mice eat cheese". After learning these statements we should be able to answer questions like "What do cats eat?" and "What do mice eat?" Simple associative learning leads to the connections of fig. 16.13 between the words.

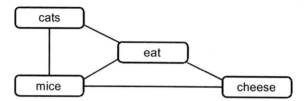

Fig. 16.13. The simple associative connections from the sentences "Cats
eat mice" and "Mice eat cheese"

If we ask "what do cats eat" and in doing so activate in
fig.16.13 the words "cats" and "eat", the word "mice" would be evoked
most strongly, by two associative connections, one from "cats" and the
other from "eat". This would also be the correct answer. So far so good,
but if we ask, "what do mice eat" and activate the words "mice" and
"eat" unfortunate things happen. The words "cats" and "cheese" will be
evoked with equal strength, by two associative connections each and
ambiguity follows. We know that mice eat cheese, but do they eat cats,
too? Of course not, mice do not eat cats. We can see that the simple
associative connections here are not sufficient to represent the
relationships correctly. In English language the ambiguity that is
involved here is removed by the word order. The question is: How
could word order be implemented in associative systems? The first idea
might be; let the system memorize complete sentences as such and
reproduce these when something is asked. Thus the question "what do
mice eat" would reproduce the learned sentence "mice eat cheese".
However, this is not a very flexible method and cannot be a credible
solution. We do not have large collections of memorized sentences in
our heads, in fact we tend to forget sentences that we hear or read very
quickly. Yet we can retain the essential information in the sentences.
For instance from a sentence like "Hungry cats want to eat fat mice" we
may later on remember only that cats want to eat mice. Thus sentences
cannot be handled as monolith sequences, we need to be able to access
the words and relationships separately inside them. A method based on
associative connections between words would offer the required
flexibility, but at the cost of ambiguity. Could we somehow augment
the method to include word order?

There is a simple solution to this problem, one that has wider
implications and is able to shed light to general problems of linguistics,
too. In order to approach this issue properly we have first to consider
alternative syntactic devices to the word order. There are languages that
do not use word order to indicate who did what to whom. In fact in
these languages the word order may be changed at will without any

effect on the basic meaning of the sentence. Even question sentences like "Is this a book?" can be expressed without the reversed word order as used in the English language. The syntactic device used in these languages is the application of inflection, the changing of the form of words.

How does this work? Here a short lesson in Finnish language will illuminate the point. Previously we had the example sentences "Cats eat mice" and "Mice eat cheese". In Finnish these sentences are "Kissat syövät hiiriä" and "Hiiret syövät juustoa". The word order is here the same as in the English example sentences. However, the words within the Finnish sentences may be shuffled any way you like without any effect on the meaning. (Some word orders may sound strange or poetic, though.) The secret here is the inflection of the words. Actually each word in these example sentences is inflected, but for the present purpose we need only to inspect the inflection of the word "hiiri" (a mouse). The basic form or the body of the word "hiiri" is "hiir". This basic form does not ever appear alone, it always has some additional ending according to the situation. Thus, in plural, if mice are the object of action, the correct form will be "hiir" + "iä". Then again, if mice are the agent, the ones that are doing something, then the correct form will be "hiir" + "et". Thus the associative connections of the two sentences "Kissat syövät hiiriä" and "Hiiret syövät juustoa" ("Cats eat mice" and "Mice eat cheese") can be depicted as in fig. 16.14.

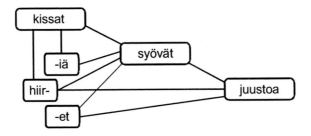

Fig. 16.14. The associative connections from the Finnish sentences "Kissat syövät hiiriä" ("Cats eat mice") and "Hiiret syövät juustoa" ("Mice eat cheese")

We will notice here that additional associative connections arise due to the inflection endings of the words. If we now ask "what do cats eat" and therefore activate in fig.16.14 the words "kissat" (cats) and "syövät" (eat), the entities "hiir-" and "-iä" will be evoked by two signals each and the word "hiiriä" (mice) will correctly result. But now to the acid test; do mice still eat cats? The activation of the entities "hiir-" and "-et" and "syövät" will lead to the evocation of the word "juustoa" (cheese) with three signals and the word "kissat" (cats) with only two signals, thus cats will no longer be eaten by mice. The entities "hiir", "-iä" and "-et" can be vertically grounded to corresponding observed entities and relationships as discussed before.

Inflection with varying word endings (or additional phonemes in front of a word) can be readily realized with distributed signal representations where each word is represented by several signals, one or more signals for each phoneme and possible additional signals for the rhythm of the word. In this way the signals that depict the body of the word will carry the general meaning, for instance the general idea of a "cat" or "mouse" and the inflection endings will carry the relationship information. This method is quite simple, something that would seem to follow naturally in associative networks that utilize distributed signal representation. Accordingly, inflected language processing can be easily implemented within the associative neural networks that have been outlined in this book.

All right then, but what has this to do with word order as a syntactic device? Surely inflection and the word order method have nothing in common and therefore must require completely different neural architectures? This is a good question and might imply that the human brain would have to wire itself differently in major ways for inflected and non-inflected languages. It would be a major breakthrough if brain imaging could prove that this is or is not the case.

But now to the revelation. Inflection and word order can be implemented within the same neural architecture after all. In associative systems almost everything can be linked to everything; in this case the relative positions of the words in a sentence may be associated with marker signals. Thus certain signals would indicate that a word has appeared early in the sentence, other signals could indicate that the word has appeared at the end of the sentence. In word order governed sentences these signals would indicate the probable agent and object and therefore would convey the same information as the inflection endings in inflected languages. Therefore the position signals could be used as "non-phonemic" inflection endings and our example sentences would generate the associative connections of fig. 16.15.

According to fig. 16.15 the question "What do mice eat" will activate the entities "mice", "[first]" and "eat". These will evoke "cheese" correctly with three signals. The question "Who eat mice" will activate "mice", "[last]" and "eat". These in turn will evoke "cats" with three signals and the correct answer is again generated. It can be seen that the required neural architecture behind the cases of figs 16.14 and 16.15 is the same; inflection and word order can be implemented with the same overall architecture!

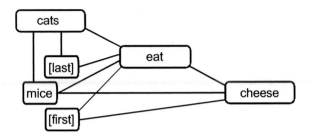

Fig. 16.15. Word order method using indicators for the relative position
of the words within sentences

In figs. 16.14 and 16.15 inflection endings and position indicators should be attached to the other words as well, this would further clarify the meaning. However, these simplified examples should already illustrate the basic principle. Nevertheless, it is useful to notice that if the sentence "cats eat mice" were given alone then the question "what do mice eat" would necessarily evoke "cats" — but with the indicator "[first]" and the complete response would be "cats [first] eat mice [last]" if the position indicators were used in the reconstruction of the word order of the response sentence. Mice would not eat cats even then.

Now to some other syntactic issues. Nested structures are common in longer sentences. Consider the following sentence: "The mouse that the cat is chasing is white". We humans can easily get the picture; there may be several mice there, one cat and of all the mice the white one is being chased by the cat. But how do we actually do this and how would an artificial cognitive system handle this kind of sentences?

The structure of the example sentence becomes more obvious if brackets are inserted as follows: "The mouse [that the cat is chasing] is white". Now it can be seen that if the words between the brackets are ignored the sentence will reduce into a simple form "the mouse is white". This can be resolved and correct associations can be made as shown before. The magic word "that" indicates that the words between the brackets are a diversion and must be handled separately; "mouse" is not to be taken as the agent that is chasing something, "white" is not to be connected to "cat". Nested sentences like these necessitate short-term-memories that are able to sustain parts of the sentences for future use. Even very rudimentary inner imagery will help here to bind related entities together. Nested structures are problematic also for us humans. Even a couple of levels of nesting may make a sentence difficult to understand, especially if the subject matter is abstract and no helping imagery can be evoked.

What do you do when you want to refer to an entity whose name you do not know or do not want to repeat? Grammar helps you here, you can use indirect reference. Instead of the actual name or noun you can use a pronoun. "It", "you", "he", "she", "this", "who", etc., are examples of pronouns. The problem with pronouns is that they do not have fixed meanings, instead the entities that they refer to depend on context.

Consider the sentences "You should see Mary. She is beautiful". The system must be able to answer "Mary" to the question "Who is beautiful?" Obviously "she" refers to the last previous female noun. Does the system have to go back, recall the previous sentence word by word and look for the last female noun? This would be rather tedious and awkward and common experience seems to show that we humans do not do it in that way. Instead I propose that two mechanisms can be used, direct association and word order. Direct association can take place if "Mary" evokes the word "she" directly allowing the connection between "Mary" and "beautiful" to be established in this way instantly.

But then, direct association is not always possible. How about "You should see Xyz. She is beautiful". In this case "Xyz" is a name that we have not heard before and thus the word cannot evoke the word "she". In this case word order can be used to indicate the referred word; "Xyz" and "she" will be associated and later on "Xyz" will be known as female.

Some pronouns are sometimes even more problematic. "We spent the night at the opera. It was fun." Which word does "it" refer to? Many of us know that opera is hardly fun, therefore "it" cannot refer to "opera". Instead what was fun was the act of spending the night with good company. Here "it" refers to an act, to a whole sentence, not to a solitary word. Yet in many cases "it" does refer to a distinct word, so what would be the syntactic rule that describes unequivocally the use of the pronoun "it"?

There is no such rule. I make here a bold generalization and claim that syntax cannot be completely separated from semantics. Every now and then vertical grounding to inner imagery and percepts of the external world is needed in order to resolve the meaning of a sentence. Therefore proper perception of external world entities and their relationships and the formation of respective inner representations are absolutely necessary. A syntactic sentence is a structured description of a given situation. If the system is not able to produce and properly bind all the required percepts then it will not be able to produce a proper verbal description either and vice versa, the system will not be able to decode the respective sentences. Artificially imposed rules alone will not solve every problem here.

Inner Speech and Machine Thoughts

We humans have this continuous inner speech that begins the moment we wake up in the morning and ceases only when we fall asleep. This inner speech is very persistent, we cannot stop it easily even for a moment. Even when we seem to succeed, we may realize that we have had it after all, for instance as a comment like "I am not thinking now". I have explained the main principles of language in the machine, now it is time to ask: Does this machine have the same kind of forced inner speech as we humans do, does this machine think in the way of inner speech and if it does, what kind of linguistic thoughts would it have?

As discussed before, percepts can evoke solitary words by vertical grounding. Horizontal connections, word-to-word associations may evoke other words. The vertical and horizontal grounding process not only gives meanings to words, it also evokes and drives inner speech.

The evoked words at the outputs of auditory association groups are distributed signal arrays that will be perceived only if they are brought back into percepts via the auditory perception process. The percepts at the output of the auditory perception process correspond to heard auditory sensations. Therefore also the inner speech that is brought back to the perception process is indeed perceived as heard speech. This inner speech may lack some of the properties of real heard speech, it has no perceived direction and there is no need to include full tonal qualities. It is also distinguished from real external speech by the lack of corresponding auditory sensor activity.

The perception/response feedback loop is actually needed for primed perception and prediction. Feedback loops in general may cause oscillations whenever the feedback signal seeds new output. Here also this phenomenon forces and sustains the flow of inner speech by causing the previous output to evoke a new output; evoked words evoke new words, etc. Therefore the inner speech is continuous as long as the neuron group output thresholds are low enough to allow output.

This inner speech involves the verbal labeling of the various sensory modality percepts; imagery, feelings, etc. In this way these are brought into the linguistic domain where they can be manipulated according to the corresponding horizontal connections, by "linguistic thinking". There can be free-running thoughts in the form of loose commentary or linguistic report about moment-to-moment situations. Inner speech as commentary may also coordinate a system's own action by questions, answers, comments, requests, commands and warnings. Goal-oriented thoughts relate to problem solving. Rudimentary free-running thoughts may arise as soon as the system has learned a modest vocabulary. Already one word can be grounded vertically but not horizontally. Horizontal grounding calls for more words and a rather large vocabulary will be needed for sophisticated horizontal word-to-word connections including syntax. The actual usage of inner speech in the form of self-directed questions, answers, etc., must be learned via examples. (When you ask yourself "why did I do this stupid thing" it is actually your mother who is asking!)

We cannot read other people's thoughts, but here as the designer of the artificial system we can probe the auditory perception process outputs and inspect the flow of inner speech. Thus we will know for sure if the machine thinks and what it thinks.

Machine thoughts are not necessarily formally different from human thoughts, especially if the machine is to learn and use human language, however the interest areas may vary as one's duties and worldly possibilities may be perceived differently. How about philosophical thoughts then? Would a machine have its own philosophy about the world and its own existence? I think so, but this we will see for sure as soon as a large enough artificial system can be built.

CHAPTER 17

INNER IMAGERY AND THINKING

Deduction and Reasoning

Deduction and reasoning involve the recovery of hidden or missing information as described in chapter 4. Deduction and reasoning processes by inner imagery call for the ability to evoke new inner representations on the basis of present ones, have them available simultaneously and detect contradictions between them. In the following I will discuss how these processes can arise within the cognitive architecture.

Simple deduction by experience involves the evocation of memories of similar cases and their outcomes. Here the representation of a possible consequence is evoked by the percept of the cause. This kind of operation is supported by the cognitive architecture directly.

Deduction by causality is an advanced form of deduction by experience. Here three valid possibilities exist:

cause is present	—	consequence is present
cause is present	—	consequence is not present
cause is not present	—	consequence is not present

Here the first case corresponds to the simple deduction by experience. True deduction by causality requires the learning of the two additional cases, a task that the cognitive system can do. Now the perception of the presence of the consequence will evoke the presence of cause, the presence of cause will evoke both the presence and non-presence of the consequence. Furthermore the suggestion "cause is not present — consequence is present" must be recognized as not valid. If this invalid suggestion is presented to the cognitive system, it will not find a matching case among the associatively evoked three valid cases and mismatch will be generated; the case will be seen as not valid.

Deduction by exclusion is based on the impossibility of the excluded alternative. The general rule is: A is either B or C. If A is fixed to be B then the proposition "A is C" is contradictory. The dollar bill is green, we can have a mental image of that. The statement "dollar is red" leads to mismatch between our existing image of the dollar bill and the proposed one. As a result mismatch emerges and a corresponding word like "no" can be evoked. This process is also supported by the cognitive architecture.

An object cannot be at two locations at the same time. Visual perception must use this basic rule inherently. A moving object must retain its exclusive identity even under noisy conditions where false percepts of the same objects might arise at different locations. This feat can be accomplished by the basic nature of distributed signal representation. Signals that depict the location of an entity can only represent one location at a time due to "winner-takes-all" threshold operation. Therefore a proposition that an object would reside simultaneously at another location will lead to mismatch state. This is another mechanism that can be used for inner imagery based deduction by exclusion.

Formal reasoning can be considered as processes where the presentation of premises, the entities and their relationships there lead to new associative connections between these entities. Formal reasoning can be performed entirely in the linguistic domain, between the word representations, as no actual meanings for the words are needed. Let's consider the sentences "Socrates is human" and "All humans are mortal" as an example, fig. 17.1.

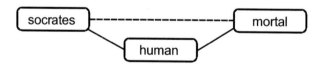

Fig. 17.1. The connections from "Socrates is human" and "All humans are mortal"

The premises are connected by the common property "human". This allows the associative connection of the information of the premise sentences within the cognitive system and the conclusion follows as a result of a new association between the entities. Thus the conclusion "Socrates is mortal" is depicted by the associative connection between the entities "Socrates" and "mortal".

Likewise the information in the sentences "No man is an animal" and "All lions are animals" can be depicted as in fig. 17.2.

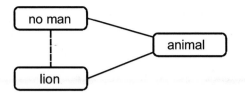

Fig. 17.2. The connections from "No man is an animal" and "All lions are animals"

Here again the premise sentences are connected via the entity "animal" and the conclusion "no man — lion" is presented by the respective new link.

Reasoning with more complicated relationships goes in a similar fashion. Let's consider the example premises "Gold is more expensive than silver" and "Silver is more expensive than bronze".

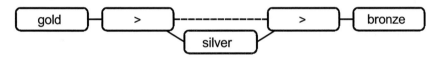

Fig. 17.3. The connections from "Gold is more expensive than silver" and "Silver is more expensive than bronze"

Here the premise sentences are connected via the entity "silver". Now we get "Gold is more expensive than (more expensive than) bronze". Had silver been less expensive than bronze then the result would have been: "Gold is more expensive than less expensive than bronze" and a mismatch would have been detected, no proper conclusion would follow.

Correct conclusions are possible because the information is there. However incorrect conclusions are also possible and here the system must learn and recognize patterns that are not valid, a feat where we humans too often fail.

Reasoning with incomplete information involves the evocation of different inner representations that relate to the given problem, detection of contradictions between them and the required conditions, evaluation of their emotional significance and the rejection of alternatives with foreseen unsatisfactory outcomes. The associative processes evoke the most familiar alternatives first. This is an inherent shortcut that, when successful, limits the scope of search. However, absolutely correct outcomes cannot be guaranteed, but then this is also the case with human reasoning.

Intelligence and Creativity

In the section on "Intelligence" (chapter 5) I concluded that intelligence refers to the ability to respond properly in new situations where old rules do not specify a suitable response. Intelligent handling of a situation calls for the detection of the requirements of the problem and its relation to experience: what is the same, what is common, what is different, what is changing, what is missing, what relationships are involved. Intelligence involves also the ability to imagine different possibilities and combinations of successful solutions to similar problems.

Within the artificial cognitive system the requirements of intelligence are met by associative processing of distributed signal representations. The distributed signal representation method represents the properties of an entity separately and allows therefore the modification of each property independently of the others. Thus an entity can be imagined to suit the requirements; evoke a known entity, make it smaller, larger, deformed, whatever is needed. The distributed signal representation allows also flexible classification by properties. Entities may be evoked and recalled by the required property. The associative operation and Winner-Takes-All threshold give the closest available response to the evoking representations of the problem's requirements. These responses are not evoked from nothing though, experience in the form of collections of related responses is needed. Thus candidate responses can be produced by associative processing, and additional mechanisms are needed for the evaluation of these responses, their actual fitness for the situation. The artificial cognitive system has these kinds of mechanisms, namely match/mismatch/novelty detection, match/mismatch pleasure/displeasure and emotional criteria, the learned and associated reward/punishment values.

Does this machine support sudden insights? As a response to a problem the associative process evokes simultaneously several responses, chains of distributed signal representations. Some of these responses are evoked only weakly, others more strongly. Only the strongest response will make its way to the feedback loop and will thus be perceived. The perceived response is matched against the requirements of the situation, if it matches, a solution has been found. Sometimes the perceived response is not a proper solution and has to be rejected. The weaker responses that are not perceived are still latent in the neuron groups. Only when the strongest response has faded away do weaker responses have their chance as the threshold is gradually lowered. This may take some time but eventually the strongest of the weaker responses will win and be suddenly perceived. If this response matches the situation, a solution has been found, superficially without perceived intermediate steps. This is the sudden insight phenomenon.

Mental Arithmetic — An Exercise in Inner Imagery

An essential part of cognition is the ability to learn algorithmic operations, tasks that can be executed by following stepwise rules. One example is mental arithmetic, the mental execution of a given arithmetic operation without the aid of pen and paper or other physical aids. This can be done by inner imagery, visualizing the task as if it were done with pen and paper.

Inner imagery based mental arithmetic calls for visual memory, visual attention, sequential memory capacity and associative memory capacity for the elementary arithmetic knowledge. This elementary knowledge would consist of the rules like $1 + 1 = 2$, $1 + 2 = 3$, etc. Sequential information would consist of the rules how to direct visual attention (gaze direction) and pinpoint the numbers of the calculation that are needed at that moment. Gaze direction is a perceivable entity and therefore it is possible to associate visual imagery with different gaze direction percepts, i.e. positions in the visual field. This imagery would be accessible by activating the related gaze direction signal. This mechanism constitutes a short-term visual memory. Visual memory would contain the numbers at correct relative positions and would allow the access of these numbers by visual attention. Let's consider an example "eleven plus twelve = ?". This would be executed with the following steps.

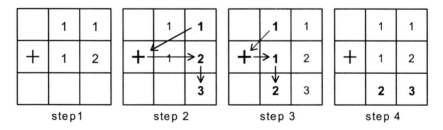

Fig. 17.4. The steps of mental addition

Step 1. The sentence "eleven plus twelve" evokes the figures "1 1", "1 2" and "+" and a visual scan sequence that associates a visual position to these figures. Thereafter these figures will be evoked when the virtual gaze direction points to the respective visual position.

Step 2. Starting from rightmost upper corner the learned visual scan sequence that is associated to addition is applied to give the sequence "1 + 2". This will evoke the figure "3". The visual position below "1" and "2" is associated to the figure "3".

Step 3. Learned visual scan sequence is applied to give the sequence "1 + 1". This will evoke the figure "2". The visual position below "1" and "1" is associated to the figure "2".

Step 4. Another visual scan sequence is applied to give the final answer "2 3".

It can be seen that mental addition involves the interplay of visual scan sequences and memorized rules like "1 + 2 = 3" which themselves are also associative sequences. Also visual memory positions that are associated to the gaze directions are needed. It must also be possible to temporarily allocate figures to these positions, sustain this imagery and reset it after the completion of the task. These requirements can be met by the cognitive architecture.

Mental arithmetic done in this way is symbolic processing. The symbols like "+", "–", "=" designate rules, series of actions to be executed. Computers and calculators utilize also symbolic processing, so what is the difference here? Consider the following example: 43263 + 768934 = ? Did you calculate this already? Good for you if you did, but somehow I believe that you did not bother to. However, whenever you enter these symbols into a calculator the machine has no choice but execute the calculation, the definition of these symbols in relation to the design of the machine forces that. We humans have a choice, we can execute symbolic processing tasks if we please, if we are motivated and do not have more urgent things to do. There is nothing in the symbols and their relation to our brains that would force us to execute the task right then and there. Likewise learned symbolic processes like mental arithmetic do not constitute forced ways of operation to the cognitive machine. The machine may or may not execute symbolic processing tasks depending on its attentional and motivational state at the given moment.

Mental arithmetic ties up lot of short-term memory and attention and is susceptible to disturbances. It is awkward whether done by humans or cognitive machines. Therefore it would be more practical to integrate an associatively accessible digital mathematics processor into the cognitive machine. Then mathematical calculations would be performed by this mathematics processor and the results would be available to the cognitive system whenever these might be needed, while the cognitive system itself could freely exercise its attention on other things.

CHAPTER 18

MACHINE CONSCIOUSNESS

Ghost in the Machine

Can a machine have an immaterial flow of mental content? What is involved in this question? Earlier in this book I have presented the basic problems of consciousness and have reviewed briefly some published models of consciousness and conscious systems. Thereafter I have introduced a novel architecture for artificial cognitive machines with the flow of inner speech and imagery. Now it is time to ask: Does this architecture offer any new insights into the quest for consciousness, would a machine with this architecture be conscious and what would this machine consciousness be like.

The first problem to be tackled is the material mind–body effect. We humans perceive our thoughts and consciousness as immaterial; as mental content only, without any perception of underlying material brain processes. However, according to materialism this is an apparent state of affairs only, not the reality as the thought process is completely material. This perceptual phenomenon is the materialistic mind–body effect as I have discussed before.

Any machine that is claimed to be conscious should also reproduce the materialistic mind–body effect. This is a fundamental question and I have not seen any plausible attempts towards an answer before. Now I am proposing one here. I am proposing that a machine with the architecture outlined here does indeed reproduce the materialistic mind–body effect; in fact it is not able to do otherwise.

The cognitive machine acquires information via perception processes as described before. This information is represented by distributed signal arrays. In the machine these signals are in the form of electrical signals; voltage levels at various transistor inputs and outputs.

By the circuit arrangement these signals are in causal connection to corresponding sensor outputs. In this way the basic meaning of each signal is fixed to a detected feature of the outside world. If we could ask the machine whether a certain signal would refer to the states of those transistors that carry it or to an external world event that causally generates the signal via a sensor, the answer would be: "Signals, what signals? I only perceive properties of the external world, I do not perceive any signals!" Thus for the machine the signals would be representations of external stimuli; detached from the material substrate that carry them. The machine would not perceive the material basis; transistors and circuit elements that carry the signals, these would be *transparent* to the machine. Circuit transparency is by no means unknown in electronics, in fact in many applications it is a desired feature. Radio, television, telephones all contain the principle of transparency. The operation of the distributed signals in the cognitive machine can be compared to radio transmission where a carrier signal is modulated to carry the actual audio signal. The carrier wave is what is received, yet what is detected is the modulation, the actual sound signal that is in causal connection to the original physical sound via a microphone. We do not hear the carrier signal even though without it there would be no music. Thus it is possible to perceive carried information without the perception of the material basis of the carrier. However, any imperfections in transparency appear as distortion and may give a clue about the underlying material carrier mechanism.

Fair enough perhaps, but in this case the information is to be carried to the machine itself, not to an external observer, does this transparency principle still apply? Yes, it does, it does not matter where the output is directed to, it is still devoid of information about the carrying machinery.

But how about abstract thoughts then? Surely an abstract idea is not grounded to the external world in any way, what would be the meaning of the signals that carry these abstract ideas? It was stated earlier that the causal connection to an external stimulus causes the signal to be perceived as a representation of that stimulus; if an abstract idea has no such connection then wouldn't the corresponding signals be perceived as what they really are, material states of the machine? We humans do have abstract thoughts and ideas and there is no reason why a machine could not have these, too. However, even our abstract ideas appear as words or mental imagery. Words, even abstract ones, are in themselves auditory signal patterns and are therefore representations of sensory stimuli. Mental imagery is again based on visual signal patterns, which also are representations of sensory stimuli. In this way abstract idea signals are also connected to external world and are thus carried as signal modulation without any information about the material carrier substrate. These representations of auditory and visual stimuli are not themselves the abstract meaning that they depict, the abstract meanings arise from associative connections to other words, images and groups of these as discussed before.

But please note, the principle of signal transparency relates only to the invisibility of the actual cognitive machinery and carrier mechanism, it does not contain the claim that external world properties were carried into the machine exactly as they are. No, this is not claimed. What the machine perceives is determined by the sensor properties and preprocessing even though the sensory processes themselves are not perceived.

We can see that the principles of transparency are realized in the cognitive architecture. The signals carry information about the sensed external world features as on/off modulation. The various pieces of information interact within the machine via the material carrier medium; the actual transistors that the artificial neurons are made of. What the machine "sees" is the carried modulation, the seemingly immaterial representations of external stimuli and not the material states of the carrier medium. *Thus the material mind–body effect does take place here and we have created the ghost in the machine*; a flow of seemingly immaterial mental content that the machine, if otherwise complex enough, could report as such.

This explanation of the material mind–body effect will apply to the human brain as well. Signal and carrier transparency is present also in human cognition. We perceive things that are out there instead of the actual neural sensory points-of-origin of these percepts; neural activation caused by images on the retina and vibrations of the eardrums, etc. For instance, it is known quite exactly how the information is carried from the eyes to the brain by neural signals. However, we do not perceive these neural firings even though we know for sure that they are there, instead we perceive the carried information. This same principle of carrier transparency is operating also inside the brain. The apparent immateriality of our thoughts is caused by omission; the inability to perceive the actual signals and machinery that carry the perceived information. This is not a result of any intricate steps of evolution, this is the simplest state of affairs.

Conscious Machine Awareness

What is involved in conscious perception? The cognitive architecture provides the perception process that is able, as discussed before, to produce mental content that is devoid of information about the carrier medium. Thus we can have the flow of "immaterial representations", but the question remains: How does the machine become aware of these representations? To whom are these presented? Do we still need a higher level observer and supervisor, a discrete machine self? What would be the difference between conscious and non-conscious perception?

It is obvious that artificial systems may acquire information about their environment by, say, a video camera and yet be patently without conscious experience. No matter how flawlessly camcorders record what is going on they are still definitely not conscious. So what is the difference between the image captured by a video camera and conscious visual perception? First of all we have to ask: For whom does a video camera capture the image? The answer is: Not for itself, but for an external observer. The captured image has no meaning for the video camera, the image does not evoke any memories, it does not suggest any possibilities for action, etc., no objects are even detected within the image. The image is only an array of numeric values that are generated by the varying light intensity across the image.

The cognitive machine is different as it utilizes perception processes that capture sensory information for the machine itself. Visual objects are detected, these evoke emotional significance and a large number of associations, possibilities for action, etc. The detected objects have a meaning that is subjective from the machine's point of view. The mere detection of these objects, the act of observation may constitute a part of the machine's personal history.

However, we know that in humans not all percepts reach consciousness even though they may affect our behavior, therefore the perception process alone is not sufficient to explain consciousness either in humans or in machines. Therefore we must investigate the difference between conscious and non-conscious perception. What happens to those signals that eventually reach consciousness? It would be easy to propose that signals become conscious when they reach certain special neuron groups. Should I thus define machine consciousness by labeling the activity of some neuron groups or the neuron groups themselves as conscious? After all this is what the search for the seat of consciousness in the human brain is about. Yes indeed, it would be easy to reproduce the block diagram of my cognitive machine here once more and fix a "consciousness" label here or there and consider the problem solved. However, I think that you, my respected reader, would not be impressed or satisfied and frankly, neither would I. Therefore a closer look at conscious and non-conscious processes is needed.

Contrastive analysis of events performed consciously and non-consciously may give the desired insight into the nature of consciousness. Now that we have the framework of the artificial cognitive architecture this kind of analysis can be done quite easily, for instance with the following examples.

What happens when the machine is reading something consciously? Within the cognitive architecture conscious reading or "reading with thought" would involve the understanding of the read text, the evocation of related imagery, emotional significance, etc. This in turn would necessitate the associative cross-connection and looping of related modalities, mainly the auditory and visual ones and at least short term learning of relevant associative connections in the style of the multimodal model of language implementation. This is depicted in the simplified fig. 18.1 where the connections between the auditory and visual modalities only are shown.

In fig. 18.1 text is sensed with the visual sensors and this visual percept is used to associatively evoke respective flow of words at the auditory loop. This flow of inner words evokes then the respective inner imagery at the visual loop. The inner words also evoke other inner words and sentences, the related knowledge in linguistic form. Cross-connections within auditory neuron groups and between other modalities allow the build-up of temporary and permanent episodic and declarative memories. Also emotional evaluation will take place and emotions may be evoked even though these are not depicted in fig. 18.1. Note that the perception of the external text and inner imagery occupy the same modality and interference between these must be avoided. This may be effected by temporal interleaving or by using different visual positions for the sensed text and evoked inner imagery, in the same way as in the case of mental arithmetic.

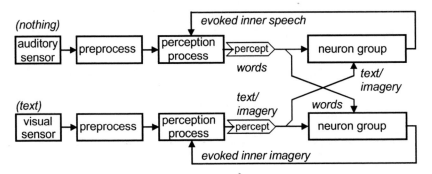

Fig. 18.1. Conscious reading or "reading with thought"

This grounding of meaning for imagery and linguistic

knowledge, evaluation of significance and the build-up of memories allows the machine to recall and understand what has been read, it allows it to answer questions about the subject matter like what happened where, who did what to whom, was it good or bad, etc. It also allows reasoning and detection of contradictions in the story. It allows paraphrasing; the description of the story with machine's own words. It also allows the setting of the emotional mood of the reader.

We humans know that we have been conscious while reading because afterwards we are able to report these things to ourselves and others. The situation here is similar, the machine is able to deliver same kinds of reports.

However, those of us with little children may know that it is possible to read aloud without any understanding or subsequent memory of what has been read. While reading aloud we may think completely different matters without any idea of what is being read. Obviously we have read the text without conscious attention, in a non-conscious manner. This I call the "Bed-time story reading effect". A similar effect is also possible within the cognitive machine architecture and is depicted in fig. 18.2.

In the "Bed-time story reading effect" the text is visually sensed and preprocessed normally and a visual percept is generated. However, in this case no cross-connections to other modalities exist and accordingly any learning of associative connections, emotional evaluation or forming of reportable memories cannot take place. Therefore afterwards no report is possible about the read matter. Only a direct path from the visual domain to the motor neurons exists, allowing the wording of sensed text in a mechanical stimulus–response way. The hallmarks of conscious perception are not present.

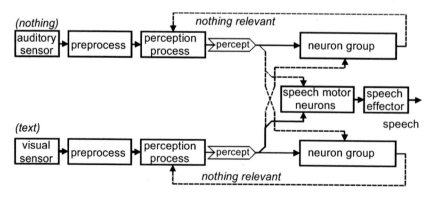

Fig. 18. 2. "Bed-time story reading effect"; non-conscious reading

As another example I present a strange brain damage related phenomenon; blindsight[23]. Patients suffering with blindsight have functional eyes but their visual cortex is damaged. Consequently these patients claim to be totally blind. However, when urged, they are able to point quite accurately towards objects that they do not see. Some patients are even able to reach out and grasp these objects if they are close enough. Yet they deny seeing anything. Numerous experiments with various patients have shown that this phenomenon is real and the patients are not just claiming to be blind due to some psychological disorder.

What could blindsight tell us about the nature of consciousness? Here obviously the patient is not conscious of what he sees even though he must see something. So, is the visual perception somehow separated from consciousness here? If so, does the damage area on the visual cortex correspond to the seat of visual awareness?

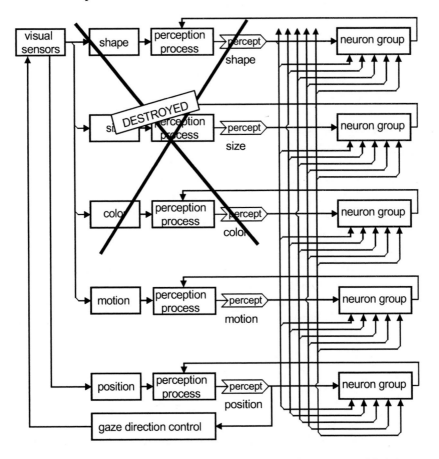

Fig. 18.3. The blindsight phenomenon produced in the artificial cognitive architecture

The phenomenon of blindsight can be reproduced within the artificial cognitive machine architecture quite easily. By doing this we will also readily understand the nature of this phenomenon and its real relation to consciousness. Fig. 18.3 depicts the visual sensory modality with the various loops that process shape, size, color, motion, distance and gaze direction information. If the shape, size and color loops are destroyed then obviously the machine will not be able to perceive these properties. However, the motion, distance and direction information remains and these percepts may be bound together and broadcast to other modalities. Thus no object with a definite shape, size and color can be perceived and reported, yet its motion and position are there, we have the blindsight phenomenon.

So, there is nothing mystical in the blindsight phenomenon, it can be completely explained by the properties of the distributed signal representation. Here the machine cannot bind shape, size and color information and report this to other modalities by broadcasting the respective percepts, it cannot become conscious of the corresponding entity. The intact motion and position information can be broadcast and the machine can act upon this information, it will be conscious of it. However, strange consciousness this is, motion and position perception without an object.

From these examples it can be concluded that the difference between conscious and non-conscious operation would be the level of active cross-connections and binding between modalities; the cross-modality reporting and learning of related associative connections and thus the establishment of episodic memories of the event. In non-conscious operation the cross-connections are minimal and the operation of the different modalities is not unified, it is not about the same topic, there is no binding. In conscious action the operation of the different modalities would be unified; the inner attention of each modality would be focused on the same topic. Thus conscious action involves focussed attention, the evaluation of meaning and emotional significance of the situation and the readiness to act upon the situation. Also the situation is reflected in the inner imagery and inner speech and can be reported and remembered.

A stimulus that is to evolve into a conscious thought appears first as a sensed representation at the input of a sensory modality or as an internally evoked representation at the various neuron groups. This representation has to fight its way through various winner-takes-all thresholds to become a percept within its own sensory modality. Percepts are broadcast to all sensory modality perception/response modules, thus this newborn percept has its challenge heard by the other modules. If the intensity of this percept is high enough or the receiving modules are not busy and therefore have low thresholds, the new percept will be received and it will evoke corresponding percepts at the receiving modules. In this way the original stimulus may eventually make all modules focus their attention to the same topic. For instance a seen object evokes its meaning, what the object is, whether it is good, bad or threatening, etc. Also the possibilities for action, what to do with it, how to handle it are evoked. Thus the original stimulus becomes the focus of attention within many or all perception/response modules and, so to say, the global inner attention is focused on the one stimulus. The percepts that are generated by the original stimulus enter also the various short-term memories and are therefore associated with any subsequent percepts. Thus the percept of the original stimulus becomes a part of the episodic memory so that it can be recalled subsequently. Now the percept of the original stimulus has all the hallmarks of a conscious percept. It has global reach, it can be reported and acted upon in the terms of other modules, its emotional significance is evaluated, it can be remembered. However, this will all take time; there will be a certain time delay between the onset of the initiating activity and the final "conscious" state, something similar to what was revealed in Libet's experiments[53]. In artificial electronic systems this delay may be very short, much shorter than in the human brain.

All these percepts that are evoked by the original stimulus may also evoke further representations that in turn will have to fight their way into winning percepts, into consciousness. Thus a kind of associative chain reaction takes place and the flow of inner representations continue; the flow of the contents of consciousness.

In the framework of this cognitive architecture the contents of consciousness are the final results of ordinary associative processes, the winning ones that engage most modalities and are available later on for retrospection.

Therefore, *consciousness is not an observer, agent or supervisor, instead it is a style and way of operation, characterized by distributed signal representation, perception process, cross-modality reporting and availability for retrospection. There is no need for special "conscious neurons", conscious matter or a special seat of consciousness.* There is no discrete machine supervisor self, the supervision is distributed in the machine. While digesting this it is useful to realize the distinction between *the contents of consciousness* and *consciousness as a process.* The above definition relates to the latter.

Thus this operational mode that has the hallmarks of consciousness appears when the numerous perception/response modules of the system are operating in unison, focused on the same topic. However, this consciousness is expensive, it engages most of the machine's perceptive, memory making and evaluation capacity on one topic only. It slows down the generation of responses as the "opinion" of all modules has to be accounted for and this causes delay. Sometimes "non-expert opinions" from some modalities might even deteriorate the final response. In many cases each module might very well be able to generate a proper response automatically, on its own. This could be a fast response, too, but no report or retrospection in terms of other modalities would be available later on. Everyday experience would seem to confirm the usefulness of this kind of automatic operation in many cases, for instance in the game of darts or sharp-shooting and in sports in general.

It should be easy to see that consciousness, when understood in this way, is intrinsically present in various degrees in cognitive systems that utilize cognitive architectures and the representation of information similar to what I have proposed here.

Self-conscious Machines

Self-consciousness may be thought to be the trickiest part of conscious experience. How can one be aware of one's own existence? How can one perceive one's own thoughts? How could a machine become aware of its own existence?

Obviously the awareness of one's existence as an individual material entity calls for the perception of one's material body that is distinct from the environment. Here also sensory percepts that give rise to the vantage point of view are crucial.

The awareness of the machine body can be achieved via suitable body sensors and the perception process in the same way as the awareness of an external environment. However, the machine must make a distinction between the sensory information originating outside the body and that originating within the body. It is assumed that the sensory signals do not carry any information about their origin except by the rigid wiring. Thus the point-of-origination would only be given by cross-associations to other signals. For instance a visually perceived nearby point may be associated with sensed manipulator (hand) position and have its visually perceived position and point-of-origination fixed in this way. Also when the machine touches itself, sensory information will be simultaneously acquired from the touching and touched part of the body. Moreover, the perceived origination points of signals that are generated by body sensors are carried with the machine when it moves around while the origination points of externally generated signals are not. These cues are readily available and can be used by the artificial cognitive machine and its associative process to create the required vantage point effect.

Signals from various body sensors constitute the "body image" of the system. The cognitive architecture can associate signals from the various sensory modalities to each other while making episodic memories. Therefore also the body sensor signals are associated with everything that happens to the machine, what kind of events it has gone through, where it has been, what physical tasks it has performed. Thus the episodic memories that are formed share a common nominator; the ever-present body, the bodily self. This is the signal complex to which the concept and word "I" can be associated initially. Eventually this signal complex will also acquire associations with the machine's will; needs and strategies to fulfill them, etc. In this way "I" will contain not only the personal history but also the style of operation, the way in which the machine pursues its goals, I might even say its personality.

However, the awareness of the machine's own body, the vantage point-of-view and self-image are only half of the self-consciousness story. Perhaps even more crucial is the requirement of the machine's ability to perceive its own mental content, the flow of the inner speech and inner imagery, the "movie-in-the-brain", memories and personal history and, still more importantly, the ability to report the existence and ownership of this mental content to the machine itself and others, to be able to have "thoughts about thoughts".

How can the awareness and reporting of mental content take place? I proposed earlier that a conscious percept has global reach, can be reported and acted upon in the terms of other modules, has its emotional significance evaluated and can be remembered at least for awhile. I also stated that any signal patterns that reach the percept points in the architecture may become conscious in this way. For instance the inner speech that flows in the auditory modality may report what is happening elsewhere in the system, what kind of inner imagery is flowing, what the other sensory modalities are perceiving. All this is enabled by cross-modality connections that allow the broadcasting of other sensory modality percepts to the auditory modality. The reporting of inner imagery in terms of inner speech is considered in fig. 18.4.

In fig. 18.4 inner imagery percepts are broadcast and reported to the auditory modality association neurons where they evoke respective linguistic expressions. In this way a linguistic report of the contents of the inner imagery may be given. However, full awareness of the flow of the inner imagery necessitates not only the perception of its content matter but also the perception of a generalized concept; the concept of inner imagery. The system must be able to report the existence of inner imagery as such, as a flow of imagery that is created by the system itself. This concept is not necessarily different from other generalized concepts like "color" and may arise from the activity of visual percept signals in general. The absence of corresponding visual sensor activity will show that these signals do not have external origin, instead they have been evoked internally via the feedback loop. Thus, the system may label this kind of visual percept activity as "inner imagery" in general and so be able to produce linguistic reports like "I have inner imagery" or the equivalent depending on the available vocabulary. In the same way the auditory system may make a linguistic report about the percepts from all the other sensory modality units.

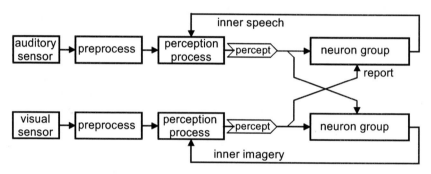

Fig. 18.4. Reporting of inner imagery

Here the auditory system acts as an outside observer for the other modalities. The observation of the flow of inner speech would also call for an outside observer that could reflect linguistically the flow of inner speech, to have "thoughts about thoughts". Could the auditory system observe itself in this way?

In this architecture the auditory system as such is not able to report the flow of inner speech to itself directly, as it cannot act as an outside observer for its own flow of percepts. However, the auditory system can achieve secondary access to its linguistic percepts if these are broadcast to other sensory modalities like the visual sensory modality and have their respective percepts, the evoked meanings returned to the auditory modality neuron group. In this way the auditory system could have a reflection of the flow of its own percepts.

However, not all thoughts are grounded in this way to inner imagery or other sensory modality percepts and therefore cannot be reported indirectly in this way. Now the problem is: What mechanism could mirror and report the flow of inner speech to the auditory modality so that there could be linguistic thoughts about thoughts?

Where is the reporting process now when the system is thinking abstract thoughts without any reference outside the auditory modality? Are thoughts about thoughts impossible here and does this system remain without consciousness of its own thoughts?

This problem might be solved by using some intricate arrangement of short-term memories. A more elegant solution can be found with the realization that there must be not one but two loops that carry inner speech. The first loop is the perception loop within the auditory sensory modality and the other is the speech motor neuron loop. The system produces sounds by audio effectors and these are driven by the speech motor neuron group as described previously. It is useful to use the general perception/response module architecture with the associative neuron group unit for the speech motor neuron group, too, so that it can produce speech sequences also on its own. For that purpose sensors that sense the speech motor neuron outputs are needed. Now we will have another percept point, one that indicates the output state of the speech motor neurons. Thus there will be two cross-connected loops, the auditory loop driving the motor loop and the motor loop reporting back to the auditory loop. Both loops carry and temporarily store inner speech. Inner speech is no longer only "heard speech" it is also silent "spoken speech"! Now we have the required mechanism that can report inner speech in terms of inner speech and thus form thoughts about thoughts.

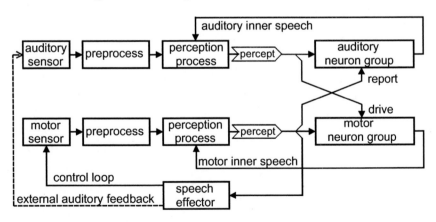

Fig. 18.5. The two loops of inner speech

Again, the reporting of the flow of inner speech as such necessitates the generation and perception of a generalized concept; the concept of inner speech. This can initially be a label, a categorical name given to the motor inner speech percepts.

Would this two-loop model apply also to humans? There is one phenomenon that would seem to support this. This is "the song in the head effect", the persistent song or melody that sometimes accompanies our thoughts. A melody sometimes with even lyrics may circulate in our mind while we think of something else and every now and then we realize that this melody has been running on there all the time. (As a child using this effect I tried to learn to think two streams of thoughts simultaneously in order to double my thinking power, however there is proof that it did not work out.) Obviously separate loops would be needed for our actual inner speech and this melody. The speech motor feedback loop could sustain the melody while leaving the auditory feedback loop free for inner speech thoughts. Indeed, PET-scan investigations would seem to confirm that both auditory perception and motor areas in the brain are activated when auditory imagery like a song in the head is perceived[42]. Thus thinking in the form of silent inner speech would indeed involve the activation of two separate but connected areas, the auditory perception area and the motor area. Whether these areas constitute loop architectures as proposed here remains to be seen.

The Ultimate Test for Thinking Machines

Does a machine think? Is it conscious? Alan Turing considered thinking as the production of mental results. Thus a machine would think if the results that it produced were not distinguishable from those produced by human mental efforts. Likewise we might infer that if a machine behaves like a conscious being then it must be conscious. However, this line of reasoning is not without problems. Similar results and behavior do not necessarily guarantee that these are the products of similar processes. A calculator can produce a result that can also be produced by human thinking, yet we do not consider a calculator as a thinking system. I have discussed these problems already earlier in this book and have rejected the principle of artificial intelligence via the imitation of results. I suppose this conclusion is accepted nowadays by many. Thus the comparison of results and behavior is not sufficient and the Turing test should be replaced by another, more fitting one. This test must consider and compare also the inner processes of the brain and the machine. Thinking and consciousness have some distinct properties like the perceived mental content; the flow of inner imagery and inner speech. These in themselves are not yet all-inclusive indicators of consciousness but may be considered as necessary ones. A system cannot be aware of its mental content if it does not have any and thus a basic requirement of consciousness would remain missing.

Thus we need a test that shows that the machine has the flow of thoughts and it is aware of them. We can determine the existence of the flow of inner speech and imagery from the architecture of the machine, we can also detect them in actual working hardware. What remains is the indication of the machine's own awareness of these. There we may have to rely on the machine's own report. If the machine is able to report that it has inner imagery and inner speech and can describe the contents of these or can even produce a conclusion "I think — therefore I exist" then it may be deemed that the machine is self-conscious. Even so there would be problems. Obviously this kind of a report would only count as a proof of self-consciousness if it could be seen that the system is producing it meaningfully. We would have to know that the machine does have the flow of inner imagery and inner speech and is indeed able to perceive these as such and can recognize them as its own products, not percepts of external stimuli. We would also have to know that the machine does have the concepts like "I", "to have" and "inner imagery". The mere reproduction of blindly learned strings of words like "I have inner imagery" would not count as a proof here. If we can determine these limiting factors from the operational principles of the machine then the proposed test will be feasible. This, I think, will be the case.

CHAPTER 19

Consciousness, Technology and Final Questions

Towards the Technology of Mind

In this book I have presented basic questions about cognition and consciousness and have outlined the nature of cognitive machines; ones that are able to perceive the world in a similar way to us, ones that have a similar vantage point of view to us, ones that have similar flow of inner speech and imagery, ones that have emotions, ones in short that are conscious. I have also sketched a possible way to construct these machines.

My approach is not purely theoretical, I have also developed experimental chips that are suitable for the implementation of the proposed cognitive architecture.

I am not alone here. Others are doing the same thing, not necessarily in the same way, but with the same goal in mind. This is a tall order though. The road is not a smooth one and there are those along the way who insist that the route is impassable. Hail, rain and shame will surely wait for the foolhardy.

However, new technology will arise. Cheap common sense in a chip will be in demand. Those who master the new technology of artificial cognition and consciousness will reap magnificent profits in growing markets. There will be new applications and products, ones never seen before. Some of them will be trivial, others quite unexpected.

Will machine consciousness, implemented on a chip, resolve the philosophical questions about mind and consciousness for once and all? No doubt, many questions will be clarified and become trivial as we become familiar with the consciousness of the machine. However, many questions may still remain open. Will machine consciousness really be similar to our own consciousness or will it only remain functionally equivalent to it? Does a machine really feel pain when it says so and acts like being in pain? Are these some of the questions that will remain open or will there be a simple answer to these? And what about humanity? What about culture? Will our views about these be changed? Will the machines have their own culture different from ours? Will they have their own philosophy? Will they have their own music? Will they have their own joys, sorrows, secret desires, perhaps even existentialist suffering? And finally, the most important question of them all: Will they be telling funny jokes about us humans, jokes that we do not even understand? I think things might get very interesting.

Skillful Robots

What weighs about one kilogram, sits on the driver's seat in a Formula One car instead of a driver and wins every race without human intervention? Driving a Formula One car with speeds exceeding 300 km/h is a demanding task with no room for errors. When errors happen, cars get smashed and lives may be lost. The best drivers are able to cope with the unexpected, they may escape situations that could be fatal to less skilled drivers. So what could be better than the best of the best?

Human reactions are limited by the speed of neural impulses and neural processing that correspond to ridiculously low clock frequencies in a computer. However the brain can compensate the inherently slow operation by its massive parallel way of processing. However, if we had a brain that were both fast and parallel then we would have the ultimate processor.

The processing speed of the artificial cognitive system is not limited by biological constraints. The speed of electrical signals approach to that of the light, the response time of individual artificial neurons can be made extremely short. The artificial cognitive system is also vastly parallel. Thus, other things being equal an artificial cognitive system should always outperform a human competitor.

The task of a racing driver is easily described. Stay on the course, do not bump into others, drive faster than others and cross the finish line first. If you follow these steps, you will always win the race. All this is easier said than done and it takes years of training to become even an average driver. The actual skills of a Formula One driver cannot be properly described verbally and even less as a computer program.

Car driving is a complex sensorimotor skill, it involves the perception of the real moment-to-moment situations and the desired immediate future situations as well as the adjusting of proper motor responses to these. The driver must be able to plan immediate actions fast based on the ever-changing situations and execute these swiftly.

Earlier I outlined how the artificial cognitive system is able to learn and plan feedback-controlled motor actions. I have also described the "mirror neuron" mechanism that allows learning by imitation. We can utilize these properties of the cognitive system and build the ultimate Formula One driver. For that purpose the system needs sensors; visual sensors to track the road and the position of one's own vehicle and others as well as possible obstacles, speed and acceleration sensors, tyre friction sensors, etc., and also sensors that indicate the instantaneous positions of the steering wheel, gears, accelerator and brake pedals. Each of these sensors needs its own perception/response loop in the cognitive architecture. The system needs also effectors that are able to operate the steering wheel, accelerator, brakes, etc.

Next we take the installed system to the races. Initially the car is driven by one or more expert drivers and the system only observes and learns the relationships between the various percepts; the position of the accelerator and the speed of the car, the position of the steering wheel and the direction of the car, braking and acceleration at curves, etc. Later on the system itself may be allowed to drive the car, the human driver only observes how the system is doing and does corrective action when necessary. Eventually the system should be able to surpass its teachers' abilities due to its faster processing speed and better perceptual connections to the various elements of the drive. Now we have no further need for human drivers.

The beauty of all this is that once we have trained one artificial driver, we can mass-produce it just like any other electronic product as we can easily copy the contents of the system. We do not have to train every new system tediously again. Also these new systems will still be able to learn more during their course of operation. These are also expendable; if accidents happen no human life is lost. Copied skills will be cheap and useful.

Minds on Silicon

What would Albert Einstein say about the latest findings about black holes? This we can more or less easily find out. We can insert the latest numeric data into Einstein's equations and get the very same answer that Einstein would give. What would Karl Marx say about the demise of Soviet Union? Would he admit his mistake or would he point out something that has been overlooked? In this case there are no numeric data, no equations to give the answer. However, we might try to figure out the answer by scrutinizing carefully Marx's life and writings. Obviously this would be quite tedious to do, but what if we had a machine to do it? A cognitive machine along the principles that I have outlined here might do the job. In fact we could load into this machine everything we know about Marx, his writings, life and emotions in the way of episodic personal history and declarative knowledge. So eventually we would have created a virtual mind; the machine would believe itself to be Marx himself. Now we could put any question we pleased and this virtual Marx would give us his opinion, mediated by all his knowledge. The process need not to end here, this virtual person would be able to learn new things as well as associate and interpret these within the framework of his line of thought. And, make no mistake here, these virtual minds would not simply be associative dictionaries with some logical reasoning power, no, these would be cognitive systems with meanings grounded to real world so that they would indeed understand what they were talking about. They would even be conscious of their own existence.

Of course, the accuracy and fidelity of the character of these virtual persons would not be hundred percent, but yet there might be distinct educational or more probably, entertainment value. So, by collecting the writings and other information about famous people — be it Aristotle, Nietsche, Galileo, his contemporary Pope, whoever — we could create virtual minds of these people. We could even make them to debate with each other. We could even have Shakespeare write another play and then yet another. How about Beethoven or Bach composing again? Or perhaps Jobim rather?

Is this the future of television entertainment, the ultimate virtual person show? Or will these virtual persons be mass-produced, and sold to people on compact disks or via internet? Would you like to buy a virtual past US president for your companion? Would you like to have the Beatles in your pocket like a MP3 player? You would only need to command "Now say something witty", or "Now compose and play another tune for me."

Does this sound a little bit far fetched? Well then, how about your loved ones that have passed away? It is not uncommon that old people talk to their departed companions, though naturally without any response. Old photographs, videos, etc., may enliven dear memories, but what if we could now have something better? What if virtual copy minds of these beloved ones had been made while they were still alive? Then we could continue talking to them and having them to talk to us with their own voice with their best wits even though they personally had passed away.

Admittedly these virtual minds would only be copies, more or less accurate, of the original persons and the thin consciousness of such virtual minds would not be that of the original person. Would there be a way to capture the real consciousness and identity of a dead person and could we transfer the consciousness of a living person into a machine?

Consciousness and Self or Raise the Dead?

The Star Trek movie makes it look so easy. Just scan the brain and body, send the information, not the actual atoms and molecules, to another location and assemble a perfect copy with a local supply of atoms. Now the person seems to be re-created, he has all the memories that the original one used to have, he has the same emotional framework, his intentions and drives are the same, he looks and acts like the original person. He will also truly believe that he is the original person. An outside observer would also be satisfied as for all practical purposes this person is the same as before and will carry out his mission as intended. So, copy the mind, brain and body and you can restore the original person even after their destruction first time around.

Cloning is considered to be a way to make biological copies of humans. In this way we could create material duplicates of the bodies and brains of any person if a sample of this person's DNA were available. Now that we have the copy of the material brain, can we also copy the self and consciousness by implanting the person's memories into the cloned brain? The technique of implanting of artificial memories is nowadays known to psychology, so given enough time all essential personal history might be implanted by existing means. This would give us a brain with the original material structure and hardwiring plus the memories of the copied person; a physical and spiritual copy. Even the cloning of dead persons may be possible if a small amount of the person's DNA were available.

So, if this copy-theory held, why shouldn't we clone, say, Hitler or Stalin? We would raise them from small babies, implant all the personal history of their originals into their minds, all their committed cruelties and so on. Eventually these copies would become to believe in their implanted memories and would believe themselves to be the actual copied persons themselves. They would also act and behave like the originals, of course within the permitted limits that we would set for them. Then, when they reach adulthood, we would take them to court and sentence them to death. And we could do this again and again, for multiple death sentences. Justice would finally take place, wouldn't it? Even death would no longer save criminals from punishment! But would we really be punishing the original criminals? Somehow I think that this disgusting idea does not quite work.

Let's consider an old car as a metaphor. A part here, a component there must be replaced every now and then to keep the car going. If I replace the windshield today, do I still have the same car? I think so. How about replacing the engine tomorrow? Still no problem. Eventually I may have replaced each and every part in the car and I still think that the car is the same. However, if I replaced all the parts at once then obviously we would not be talking about the same car any more, I could as well take the register plates to another car of similar make. I would have broken the continuity in the identity of the car. In the same way the Stark Trek teleportation or any scan-and-copy scheme breaks the continuity of the self. You will not survive the beaming down, you will be dead. Another person will emerge down there, your copy, not you and you will have no knowledge of that, ever.

No, a perfect biological copy, even with a perfectly copied mind, is not the original even if the original were disposed of. The original self is not retained and the Star Trek captain Kirk always dies. This may not concern much those who do not utilize teleportation, but personally I would not volunteer into any beaming up. The copy-method does not transfer the self.

Can We Get into the Machine?

Diamonds and silicon chips are forever. Could we then achieve immortality by moving from our biological body into silicon based one? Our biological body is not very suitable for space travel, an artificial body with silicon brain would withstand the harsh space and endure endless journeys much better. What we need to do is to transfer our mind into the machine in order to be ready for the greatest adventure into time and space. Science fiction says that this is easy.

If we assume that each synapse in the brain represents one bit of information in digital terms then the total information content of the brain with its hundred thousand million neurons with some thousand synapses each would be around ten thousand gigabytes. The memory capacity of a present day personal computer can easily be more than 100 gigabytes which is not very far from the estimated brain capacity. Therefore in the near future it should be quite possible to load the contents of a brain into a machine as far as memory capacity is concerned.

According to the scan-and-copy theory, we only need to scan the information in one's brain and install it into a large enough computer. Thereafter this machine will posses the personal history of the original person and therefore might very well believe itself to be the actual person. Does the original person's consciousness now continue its existence in this machine, have we also transferred the self?

We have done nothing of the kind. We have not transferred the self, we have not even copied the consciousness. The only thing that can be injected into a digital computer is binary code. At best we have loaded a digital description (obviously a very large one!) of the synaptic states of the brain. This is a far cry from the conscious self, copied or original one.

People who sell cheap science fiction as true science aside, is there any serious way to approach this question? First of all we must consider what it would involve to be inside a machine.

We know that we are inside ourselves because we have this vantage point view of the world. Thus, as the first step towards migrating into a machine we would have to change the vantage point reference to one that is inside the machine. This may seem easy enough. When we play virtual reality games we seem to be somewhere else than we really are. We can move around a computer-generated virtual world that replaces our everyday one. Has our vantage point of reference changed and is this the way to eventually get inside the machine as science fiction readily suggests? No. Our perception has been distorted and we may believe to be elsewhere, but we are still the same perceiving point in space, our vantage point of reference has not changed. Our brain is still the machine that does the thinking.

So, how can we truly change the vantage point reference and the seat of mind into a machine? To do this we must have a machine that has an architecture and way of operation that parallels that of the brain. Then we might cross-connect these and eventually be able to transfer the vantage point reference and the seat of mind from the brain to the machine while preserving the personal history and continuity of self. To illustrate the point I assume here that the brain can be very roughly represented by the architecture of my artificial cognitive system, so that percept-representation points and associative input points can be identified, at least in principle. The cross-connection of the brain and the machine can now be depicted in a very simplified way as in fig. 19.1.

Here the biological senses are duplicated in the machine. Image sensors can be used for eyes, microphones for ears, etc. When the machine is connected to the brain in the way of fig. 19.1 it is like having additional perception/response modules. The brain will have additional sensory power and additional vantage point reference, which is made to coincide with that, provided by the biological sensors. It is like a blind person who suddenly recovers vision or a deaf person who regains some hearing via artificial implants. However, the machine percepts remain within the machine, the brain is only able to associate these to its existing sensory modality percepts and linguistic representations, it is able to report in terms of its own representations what the machine perceives. Still, the brain is also able to evoke representations within the machine and these again are reported to the brain. The situation is symmetrical, the machine will also have a report what the brain is perceiving and thinking. The combined brain-machine will have its common personal history stored both in the brain side and the machine side. The machine side will also learn the brain's earlier history every time memories are recalled. After a long period then the machine side would, at least in principle, contain a copy of the brain's personal history and mode of thinking.

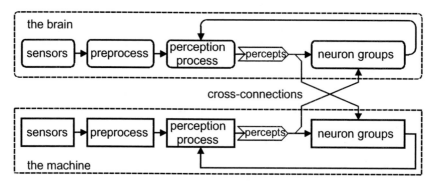

Fig. 19.1. Cross-connected brain and machine; don't try this at home

Consciousness, as I have argued before, arises in suitable systems from interaction and communication between modules. The combined brain-machine is this kind of a system and therefore would share one consciousness. The individual perception/response modules would be responsible for the flow of their specific percepts and reports in their own terms about the activity in the other modules. The removal of any of these modules would only reduce the system's capacity in that specific sense. The removal of the visual module would deprive the system of visual percepts and inner imagery, but otherwise the system would remain operational and conscious. In the same way when the brain side is disconnected after a long period of co-existence the machine side remains operative and conscious. As the sensory modalities of the biological system are duplicated in the machine side, no actual sensory capacity is lost. The self of the brain-machine system resides now in the machine and retains its continuity. The vantage point of reference relies now on the machine sensors only. The system believes to be the same brain-machine system as before albeit without the additional capacity provided by the brain. We have now completed the transfer of the self into the machine.

In practice there will be immense problems of how to make the necessary cross-connections to the brain. It would already be difficult enough to locate the proper perception points in the brain. In addition to that a very large number of new synapses would be needed for associative cross-connections at the cortex. Considering this I don't think that this kind of brain-machine connection could be done any time soon if ever. We may be able to create loose copies of our minds in the machine, but most probably we will not be able to transfer the self.

The Final Question: Being Inside

When little children ask: "Where did I come from?" you may answer: "Your mother gave birth to you." This is the practical answer to the question, but not necessarily the one that the child is after. The child has realized that there is something special in being here, to exist and to be aware of this existence, to exist as a self-conscious individual. I can remember this from my very early childhood, the spells of strange feeling that there is a deep mystery there and then the fleeting dizzy moment when I felt that if I only could formulate properly the grand question to which my very existence would be the proof and answer then the mystery would be solved. Is there indeed a proposition about our being here that our very existence would prove? How should we formulate this question? True, we are all born, but why are we born as ourselves, not as our sisters or brothers, or as some animals? Why were we born when we were? If we had not been born then, would we be born at some other time or would we remain non-existent forever? Are we non-existent before conception? Where and what was I before I was conceived? From where does the self come into our bodies? Where does the identity of the self come from?

To consider these questions we must trace back our existence, identity and scope of consciousness to the moment of conception and beyond.

Consciousness arises from perception, without percepts there is no consciousness. If there is no nervous system to process percepts then no consciousness can exist, therefore consciousness can only arise after conception alongside the developing nervous system. The scope of this consciousness is limited and unorganized at first and can get more organized only after birth when perceptual stimuli will become immensely more varied. Thus our consciousness as a process and our accumulated contents of consciousness, our personal memories do not go back beyond our birth.

How about our identity? Before conception we existed as an ovum from the very early age of our mother. Before that we existed as a genetic code in the DNA, as molecular instructions how to assemble ova. Now this is a tricky part. The genetic code obviously specifies only how to assemble ova in general, not how to assemble each ovum as an individual. This would seem to be the branching point where the identity of the new potential individual to be splits from the identity of the cell in division. Nevertheless in this way there is a biological continuum back to the very primordial material combination that first began to divide and multiply. This is when we got inside ourselves, this is the earliest point of our identity. This is clear if we consider this as outside observers, but alas, we are none, we are a part of the chain and doomed to ponder our position there. Perhaps the problem of our existence as a self and its identity is not a technical one, perhaps it is a psychological problem. In that case we should ask what kind of cognitive deficiency would be responsible?

Try as we may, our powers of introspection cannot reveal the material processes behind our thoughts and mind. For centuries we have taken this fact at its face value and have come to the conclusion that our mind was indeed an immaterial substance. So strong this illusion is that even today, when we are able to detect and measure actual material brain processes, there are people who wish to deny any connection between these material processes and the mind.

Will a conscious machine also, due to the apparent immaterial nature of its thoughts, reproduce Descartes' error and infer the existence of immaterial mind? Might it even reinvent Bishop Berkeley's idealism? Will the machine, too, come to the conclusion that there is something special in being here, to exist and to be aware of this existence, to exist as a self-conscious individual? Will it, prompted by these existentialist feelings, eventually ask: "Where did I come from?" We have created this machine and therefore there is no mystery about the origin of the machine's self and identity. If the machine has a conscious self, we know where it came from, we know the answer — we can kick the stone — but will the machine, from its subjective point of view and existential suffering, be satisfied with our answer? Shall we then console:

"My child, what does it matter,
you will live forever,
you will never die".

References

1. Aleksander Igor, *Impossible Minds My Neurons My Consciousness*, Imperial College Press, London, 1996
2. ibid. p. 9

3. Aleksander Igor, *How to Build a Mind*, Weidenfeld & Nicolson, Great Britain, 2000

4. Aleksander Igor, Morton Helen, *Neurons and Symbols*, Chapman & Hall London, 1993

5. Ashcraft Mark H., *Fundamentals of Cognition*, Addison Wesley Longman Inc, New York, 1998, pp. 67–94
6. ibid. pp. 75–76
7. ibid. pp. 97–197
8. ibid. p. 101
9. ibid. pp. 338–340

10. Baars Bernard J., "Understanding Subjectivity: Global Workspace Theory and the Resurrection of the Observing Self" in *Journal of Consciousness Studies*, Vol 3, No 3, 1996, pp. 211–216

11. Baars Bernard J., *In the Theater of Consciousness*, Oxford University Press Oxford New York., 1997, p. 74
12. ibid. pp. ix, 41–42

13. Barber J., Mayer D., "Evaluation of efficacy and neural mechanism of a hypnotic analgesia procedure in experimental and clinical dental pain" in *Pain*, 4, 1977, pp. 41–48

14. Brentano Franz, *Psychology from an Empirical Standpoint*, A. Rancurello, D. Terrel and L. McAlister (trans.), London, Routledge and Kegan Paul 1973

15. Browne Chris, Evans Richard, Sales Nick and Aleksander Igor (1997) "Consciousness and Neural Cognizers: A Review of Some

Recent Approaches" in *Neural Networks*, Volume 10 Number 7, October 1997 pp. 1303–1316

16. Butter Charles M., *Neuropsychology: the Study of Brain and Behaviour*, Brooks/Cole Publishing Company Belmont California, 1968, pp. 137–155
17. ibid. pp. 151–152

18. Casti John, *Five Golden Rules, Great Theories of 20th–Century Mathematics*, John Wiley & Sons, New York, 1996, pp. 137–180

19. Chalmers David J., "Facing up to the Problem of Consciousness", in *Journal of Consciousness Studies*, Vol 2, No 3, 1995, pp. 200–219

20. Chomsky, N., *Aspects of a Theory of Syntax*, Harward University Press, Cambridge, MA, 1965

21. Churchland Paul M., *Matter and Consciousness*, A Bradford Book, The MIT Press Cambridge Massachusetts, 1993, p. 51

22. Crick Francis, *The Astonishing Hypothesis*, A Touchstone Book, New York, 1994, p. 14
23. ibid. pp. 171–173
24. ibid. p. 249

25. Damasio Antonio R., "How the Brain Creates the Mind", *Scientific American*, December 1999, pp. 74–79

26. Damasio Antonio, *The Feeling of What Happens*, Vintage, London, 2000

27. Deacon Terrence, *The Symbolic Species*, Penguin Books, England, 1997, p. 104
28. ibid. p. 109

29. Dennett Daniel C., *Consciousness Explained*, Little, Brown and Company, Boston, Toronto, London, 1991 pp. 111–138
30. ibid. pp. 369–411

31. Devlin Keith, *Goodbye Descartes*, John Wiley & Sons, Inc. New York, 1997, pp. 143–184

32. Edelman Gerald M., Tononi Giulio, *Consciousness*, Penguin Books ltd, London, 2000

33. Fodor Jerry A. *The Language of Thought*, Thomas Y. Crowell, New York, 1975

34. Gillet Grant, *Representation, Meaning and Thought*, Claredon Press Oxford U.K. 1992, p. 34

35. Greenfield Susan A., *Journey to the Centers of the Mind*, W. H. Freeman and Company New York 1995, pp. 43–52

36. Gregory Richard L., *Eye and Brain*, Fifth Edition, Oxford University Press, Oxford 1998, p. 55

37. Guigon Emmanuel, Burnod Yves, "Short-Term Memory in *The Handbook of Brain Theory and Neural Networks*, Michael A. Arbib, editor, MIT press U.S.A. 1995, pp. 867–871

38. Gödel Kurt, "Über formal unentscheidbare Sätze der Principia Mathematica und verwandter Systeme, I", in *Monatshefte für Mathematica und Physics*, 38 (1931). pp. 173–198

39. Haikonen Pentti O. A., *An Artificial Cognitive Neural System Based on a Novel Neuron Structure and a Reentrant Modular Architecture with Implications to Machine Consciousness*, Dissertation for the degree of Doctor of Technology, Helsinki University of Technology, Applied Electronics Laboratory, Series B: Research Reports B4, 1999
40. ibid. pp. 81–86
41. ibid. pp. 103–141

42. Halpern Andrea R., Zatorre Robert J., "When That Tune Runs Through Your Head: A Pet Investigation of Auditory Imagery for Familiar melodies" in *Cerebral Cortex* Oct/Nov 1999 pp. 698–704

43. Hari Riitta, Lounasmaa Olli, "Neuromagnetism: Tracking the Dynamics of the Brain", in *Physics World* May 2000, pp. 33–38

44. Hinton Geoffrey E., McClelland James L., Rumelhart David E., "Distributed Representations" in *The Philosophy of Artificial Intelligence*, pp. 248–280, Margaret A. Boden, editor, Oxford University Press, New York, 1990

45. Jain Anil K., Mao Jianchang, Mohiuddin K.M., "Artificial Neural Networks: A Tutorial", in *Computer*, March 1996, pp. 31–44

46. Johnson-Laird Philip, *The Computer and the Mind*, (Second Edition), Fontana Press U.K., 1993, p. 334

47. Kodratoff Yves, *Introduction to Machine Learning*, Pitman Publishing, London,1988, pp. 86–92

48. LaBerge David, *Attentional Processing*, Harward University Press, Cambridge, Massachusetts, London England, 1995, p. 3

49. LeDoux Joseph, *The Emotional Brain*, Simon & Schuster, New York, NY, 1996, pp. 79–80
50. ibid. pp. 80–85
51. ibid. pp. 138–178

52. Lemley Brad, "Do You Love This Face", in *Discover* Vol 21 No. 2, February 2000

53. Libet Benjamin, "The neural time factor in conscious and unconscious events", in *Experimental and Theoretical Studies of Consciousness*, Wiley, Chichester (Ciba Foundation Symposium 174), 1993, pp. 123–146

54. Loftus Elizabeth F., "Creating False Memories" in *Scientific American*, September 1997, pp. 50–55

55. Luria A. R., *The Working Brain*, Basic Books, A Division of HarperCollins Publishers, 1973, pp. 256–279

56. McCulloch W. S. and Pitts W., "A Logical Calculus of Ideas Immanent in Nervous Activity", in *Bulletin of Mathematical Biophysics*, Vol. 5, 1943, pp. 115–133

57. McGurk H. & MacDonald J. W., "Hearing lips and seeing voices" in *Nature*, 264, 1976, pp. 746–748

58. Mealy G. H. "A Method for Synthesizing Sequential Circuits", *The Bell System Technical Journal*, 34, 5, pp. 1045–1079, Sept. 1955

59. Moore Brian C. J., *An Introduction to the Psychology of Hearing*, Fourth Edition, Academic Press, San Diego, California, 1997, p. 251

60. Moore, Edward F., "Gedanken-experiments on Sequential Machines", in Shannon and McCarthy (eds.) *Automata Studies*, Annals of Mathematics Studies, No 34, 1956, pp. 129–153.

61. Nairne James S., *The Adaptive Mind*, Brooks/Cole Publishing Company USA, 1997, p.188
62. ibid. pp. 209–210
63. ibid. p. 230
64. ibid. p. 234
65. ibid. pp. 239–273
66. ibid. pp. 256–268
67. ibid. pp. 268–279
68. ibid. pp. 280–282
69. ibid. p. 411
70. ibid. p. 413
71. ibid. pp. 384–417
72. ibid. pp. 488–489

73. von Neumann, John: "First Draft of a Report on the EDVAC", 30. June 1945, edited and corrected by Michael D. Godfrey, in *IEEE Annals of the History of Computing*, Vol. 15, No. 4, 1993

74. Nicholls John G., Martin A. Robert, Wallace Bruce G., *From Neuron to Brain*, third edition, Sinauer Associates, Inc Publishers Sunderland, Massachusetts, U.S.A., 1992, pp. 184–236
75. ibid. pp. 237–268
76. ibid. pp. 321–324
77. ibid. pp. 487–497
78. ibid. pp. 494–495

79. Nishitani N., Hari R., "Temporal Dynamics of Cortical Representation for Action" in *PNAS*, January 18, 2000, vol 97, no 2, pp. 913–918

80. Norman, D. A., "Toward a Theory of Memory and Attention" in *Psychological Review* 75, 1968, pp. 522–536

81. Näätänen Risto, "The role of attention in auditory information processing as revealed by event-related potentials and other brain measures of cognitive function" in *Behavioral and Brain Sciences*, vol 13, number 2, June 1990 pp. 201–233 Cambridge University Press

82. Näätänen Risto, Alho Kimmo, "Mismatch Negativity–a Unique Measure of Sensory Processing in Audition" in *Intern. J. Neuroscience* 1995, Vol. 80, pp. 317–337

83. Plutchik Robert, *Emotion: A Psychoevolutionary Synthesis*, Harper and Row, New York 1980

84. Rizzolatti G., Luppino G., Matelli M., "Grasping Movements: Visuomotor Transformations" in *The Handbook of Brain Theory and Neural Networks*, Michael A. Arbib, editor, pp. 438–441, MIT press U.S.A. 1995

85. Rolls Edmund T., "Consciousness in Neural Networks?" in *Neural Networks* Vol. 10 Number 7 October 1997 pp.1227–1240

86. Rose Steven, *The Making of Memory*, Bantam Books Great Britain, 1992, pp. 154–158

87. Rosenfield Israel, *The Strange, Familiar and Forgotten*, Picador, Great Britain, 1995, pp. 109–111

88. Schmajuk Nestor A., Cognitive Maps, in *The Handbook of Brain Theory and Neural Networks*, Michael A. Arbib, editor, pp. 197–200, MIT press U.S.A. 1995

89. Searle John R., "Minds, Brains, Programs" in *The Behavioral and Brain Sciences*, number 3, 1980, pp. 417–427, Cambridge University Press

90. Searle John R., *Minds, Brains & Science*, Penguin Books Ltd., London England, 1984

91. Searle John R., *The Mystery of Consciousness*, Granta Books, London, 1997

92. Simons Daniel J., Levin Daniel T. "Failure to Detect Changes to People during Real-world Interaction", in *Psychonomic Bulletin and Review*, vol. 4, p 644, 1998

93. Sloman Aaron, "Introduction: Models of Models of Mind" in *Proceedings of the AISB'00 Symposium on How to Design a Functioning Mind*, 17–20th April, 2000, pp. 1–9

94. Sommerhof Gerd, *Understanding Consciousness* Sage Publications, London. Thousand Oaks. New Delhi, 2000

95. Spiegel D., Albert L. H., "Naloxone fails to reverse hypnotic alleviation of chronic pain", in *Psychopharmacology*, 81 1983, pp. 140–143

96. Taylor J. G., "Neural Networks for Consciousness" in *Neural Networks*, Vol 10, Number 7, October 1997, pp. 1207–1225

97. Taylor J. G., "Constructing the Relational Mind" in *Psyche*, 4(10), June 1998, http://psyche.cs.monash.edu.au/v4/psyche-4-10-taylor.html

98. Taylor John G., *The Race for Consciousness,* A Bradford Book, The MIT Press, Cambridge, Massachusetts, London, England, 1999

99. Thompson Richard F., *The Brain*, W. H. Freeman and Company, New York, 1985, pp. 96–99
100.   ibid. pp. 189–223
101.   ibid. pp. 218–223
102.   ibid. pp. 249–252
103.   ibid. p. 339
104.   ibid. pp. 339–387

105.   Turing Alan M., "On Computable Numbers, with an Application to the Entscheidungsproblem" in *The Proceedings of the London Mathematical Society*, ser. 2, vol. 42 (1936–7), pp. 230–265

106.   Turing Alan M., "Computing Machinery and Intelligence" in *Mind* LIX, no 2236 Oct. 1950, pp. 433–460

107.   Winston Patrick Henry, *Artificial Intelligence*, Addison-Wesley Publishing Company, Reading, Massachusetts, 1984, pp. 295–314

Index

Machine Consciousness
Owen Holland (ed.)

Can a machine really be conscious? Can it have feelings? Can a machine think thoughts? Does it have a personality? How would you know? In this collection of essays we hear from an international array of computer and brain scientists who are actively working from both the machine and human ends of things to bridge the gap between the mind and the machine. They include Igor Aleksander, Susan Blackmore, Rodney Cotterill, Stan Franklin, Stevan Harnad, Ray Kurzweil, Jesse Prinz, Aaron Sloman and William Irwin Thompson.

250 pages   £14.95/$24.95   0907845_24X (pbk.)

How Could Conscious Experiences Affect Brains?
Max Velmans

In daily life we take it for granted that our minds have conscious control of our actions, at least for most of the time. But many scientists and philosophers deny that this is really the case, because there is no generally accepted theory of how the mind interacts with the body. Max Velmans presents a non-reductive solution to the problem, in which 'conscious mental control' includes 'voluntary' operations of the *preconscious* mind. On this account, biological determinism is compatible with free will.

96 pages   £9.50/$14.50   0907845_398 (pbk.)

Is the Visual World a Grand Illusion?
Alva Noë (ed.)

There is a traditional scepticism about whether the world 'out there' really is as we perceive it. A new breed of hyper-sceptics now challenges whether we even have the perceptual experience we think we have. According to these writers, perceptual consciousness is a kind of false consciousness. This view grows out of the discovery of such phenomena as change blindness and inattentional blindness. Edited by philosopher Alva Noë, contributors include psychologists Susan Blackmore, Arien Mack and Bruce Bridgeman and philosophers Daniel Dennett, Andy Clark, Jonathan Cohen and Charles Siewert.

209 pages   £14.95/$24.95   0907845_231 (pbk.)

TOCs, abstracts, sample chapters and secure ordering: **www.imprint-academic.com**

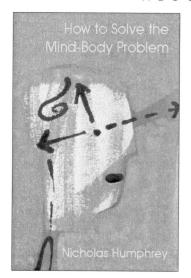

How to Solve the Mind–Body Problem
Nicholas Humphrey

The mind is the brain. Each mental state—each hope, fear, thought—can be identified with a particular physical state of the brain, without remainder. So argues Nicholas Humphrey in this readable yet scholarly essay in evolutionary psychology.

'Presents readers with a myriad of important issues.' **TICS**
'Sets out some interesting new avenues of investigation.' **Metapsychology**
112 pages £6.95/$12.50 0907845_088 (pbk.)

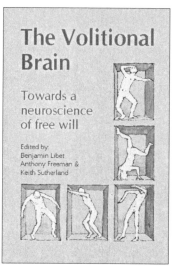

The Volitional Brain: Towards a neuroscience of free will
Benjamin Libet *et al.* (eds.)

It is widely accepted in science that the universe is a closed deterministic system in which everything can be explained by purely physical causation. And yet we all experience ourselves as free will. The scientists and philosophers in this book address this puzzling issue.

'The collection reports some fascinating empirical work on the brain activity that underlies volition.' **A.C. Grayling,** *TLS*
'A timely compilation of essays.' **Contemporary Psychology**
'Modern answers to the ancient problem of free will.' **J. Neurol. Neurosurg.**

320 pages £14.95/$24.95 0907845_118 (pbk.)

Art and the Brain
Joseph Goguen & Erik Myin (eds.)

Is it possible to take a natural science approach to art and uncover general laws of aesthetic experience, or is that taking reductionism too far? Contributors include Semir Zeki, V.S. Ramachandran, Nicholas Humphrey and Erich Harth. Discussants include Bernard Baars, Paul Bloom, Daniel Dennett, Uta Frith, Richard Gregory, Jaron Lanier, Colin Martindale, Chris McManus, Steven Mithen and Ian Tattersall.

'A penetrating neurological theory.'
Christopher Tyler, *Science*
'Distinguished exponents of their various specialisms.'
John Nash, *Interdisciplinary Science Reviews*

160 pages £14.95/$24.95 0907845_452 (pbk.)

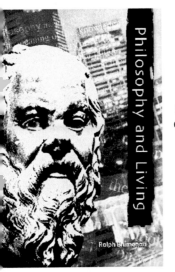
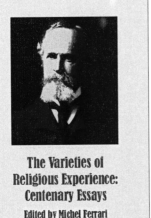

The New Idea of a University
Duke Maskell and Ian Robinson

The New Idea of a University is an entertaining and highly readable defence of the philosophy of liberal arts education and an attack on the sham that has been substituted for it. It is sure to scandalize all the friends of the present establishment and be cheered elsewhere.

'A seminal text in the battle to save quality education.' **THES**
'This wonderful book ought to make the powers that be stop and think.' **Chris Woodhead, Sunday Telegraph**
'A severe indictment of British universities' **Oxford Magazine**

208 pages   £12.95/$20   0907845_347 (pbk.)

Education! Education! Education!
Stephen Prickett & P. Erskine-Hill (eds.)

This book, subtitled 'Managerial ethics and the law of unintended consequences', criticises the new positivism in education policy, whereby education is reduced to those things that can be measured by 'objective' tests. Contributors include Libby Purves, Evan Harris, Rowan Williams, Roger Scruton, Robert Grant, Bruce Charlton and Anthony Smith.

'A call to action.' **Dom Antony Sutch, The Tablet**
'Raises issues that should be widely debated in the media and which should inform educational manifestos for the next election.' **David Lorimer, Network**

200 pages   £14.95/$24.95   0907845_363 (pbk.)

Universities: The recovery of an idea
Gordon Graham

This extended essay questions whether the recent spate of state-sponsored managerial initiatives amount to a betrayal of J.H. Newman's classic vision of the liberal arts university.

'Those who care about universities should thank Gordon Graham.' **Anthony O'Hear, Philosophy**
'It is extraordinary how much Graham has managed to say and to say so well in so short a book.' **Alasdair MacIntyre**
'Though densely and cogently argued, this book is extremely readable and indeed deserves to be widely read.' **Philosophical Quarterly**

136 pages   £8.95/$14.95   0907845_371 (pbk.)